Lecture Notes in Electrical Engineering

Volume 433

About this Series

"Lecture Notes in Electrical Engineering (LNEE)" is a book series which reports the latest research and developments in Electrical Engineering, namely:

- Communication, Networks, and Information Theory
- Computer Engineering
- Signal, Image, Speech and Information Processing
- Circuits and Systems
- Bioengineering

LNEE publishes authored monographs and contributed volumes which present cutting edge research information as well as new perspectives on classical fields, while maintaining Springer's high standards of academic excellence. Also considered for publication are lecture materials, proceedings, and other related materials of exceptionally high quality and interest. The subject matter should be original and timely, reporting the latest research and developments in all areas of electrical engineering.

The audience for the books in LNEE consists of advanced level students, researchers, and industry professionals working at the forefront of their fields. Much like Springer's other Lecture Notes series, LNEE will be distributed through Springer's print and electronic publishing channels.

More information about this series at http://www.springer.com/series/7818

S.R.S. Prabaharan · Nadia Magnenat Thalmann
V.S. Kanchana Bhaaskaran
Editors

Frontiers in Electronic Technologies

Trends and Challenges

Springer

Editors
S.R.S. Prabaharan
School of Electronics Engineering (SENSE)
VIT University
Chennai, Tamil Nadu
India

V.S. Kanchana Bhaaskaran
School of Electronics Engineering (SENSE)
VIT University
Chennai, Tamil Nadu
India

Nadia Magnenat Thalmann
Institute for Media Innovation (IMI)
Nanyang Technological University (NTU)
Singapore
Singapore

ISSN 1876-1100 ISSN 1876-1119 (electronic)
Lecture Notes in Electrical Engineering
ISBN 978-981-10-4234-8 ISBN 978-981-10-4235-5 (eBook)
DOI 10.1007/978-981-10-4235-5

Library of Congress Control Number: 2017934326

Printed on acid-free paper

This Springer imprint is published by Springer Nature
The registered company is Springer Nature Singapore Pte Ltd.
The registered company address is: 152 Beach Road, #21-01/04 Gateway East, Singapore 189721, Singapore

Preface

This volume is based on the keynote and invited lectures presented at International Conference on NextGen Electronic Technologies–Silicon to Software—ICNETS^2. It is explicit that the electronic evolution is centred on hardware and software co-design perspectives at present and beyond. The conference theme "SILICON TO SOFTWARE" relates the links between hardware (silicon and beyond) and software driven systems. ICNETS^2 is envisaged as a forum for engineers, scientists, industry personnel, postdocs and budding researchers to present their research findings and share their knowledge/experience relating to the recent advancements and current opinions in the field of Electronics and Communication Engineering. Given its importance in the arena of Electronics and Communication Engineering and Technologies, ICNETS^2 facilitates to the needs of specific topics covered in all niche aspects of relevant current scientific interests as well as offering a great leap towards futuristic goals of next generation technologies. The conference is comprised of the following six symposia covering the major aspects of Electronics and Communication Engineering domains:

Symposium A—Computational Signal Processing and Analysis
Symposium B—Intelligent Embedded Systems
Symposium C—Nanoelectronic Materials and Devices
Symposium D—Optical and Microwave Technologies
Symposium E—VLSI Design: Circuits, Systems & Applications
Symposium F—Wireless Communication Networks & Internet of Things

Considering the international standings, the organizing committee of ICNETS^2 has invited over 20 international speakers who are renowned in their relevant subject areas. Hence, their testimony has been collated in the form of their scientific findings as chapters in this book. As editors of this special volume, we are pleased to introduce the proceedings comprised of a collection of chapters contributed by invited and keynote speakers of ICNETS^2. The goal of this book is to render an up-to-date understanding on the above topics by the renowned subject experts who are invited to share their current scientific opinions and achievements in these exhilarating fields in which this international conference has been embarked on.

Thus, the book volume makes possible for budding and non-experts in these topics to gain insight into the discussed topics such as signal processing, smart robot, big data analytics, nanoelectronics, solar cell efficiency, optical wireless communication, sensors and actuators, wireless sensor networks, cognitive radio networks, microwave antennas, RF IC design, nanomaterials, information and network security and it is hoped it will possibly yield new perspectives and propensity for the better understanding in their field of studies.

Chennai, India S.R.S. Prabaharan
Singapore Nadia Magnenat Thalmann
Chennai, India V.S. Kanchana Bhaaskaran

Acknowledgements

Words fall short to express our heartfelt thanks to the authors, invited and keynote speakers, symposia co-chairs, various committee heads, and conference/symposia managers for their untiring efforts to make this conference a reality.

One of us (SRSP) wishes to submit his acknowledgement to the respected Chancellor Dr. G. Viswanathan for his blessings and support. Editors are indebted to their respective University management—Vice President Mr. Sankar Viswanathan, and Assistant Vice President Ms. Kadhambari S. Viswanathan of VIT University, Chennai and Nanyang Technological University, Singapore for their great support, enthusiasm and encouragement to organize this mega event in VIT University, Chennai, India.

Our thanks are due to Vice Chancellor and Pro-Vice Chancellor of VIT University for their support and encouragement.

SRSP wishes to record his deep sense of gratitude to all the financial and technical co-sponsors for their support without which the conference would not be a success.

Financial support rendered by Central and State government organizations in India are gratefully acknowledged. Besides, a special mention is imperative for the financial support rendered by the President Prof. B.V.R. Chowdari, MRS (Singapore) and Institute of Physics (UK).

Our special thanks go to organizing secretary, treasurer, web-page and outreach co-chairs, symposia co-chairs, sponsorship co-chair, scientific programme committee chairs, publication co-chairs, CMT co-chair, event coordinators, coordinators of industry–academia forum and all SENSE faculty members who contributed towards organization of ICNETS^2.

Editors wish to acknowledge the important contributions of Dr. Mohanaprasad in coordinating and assembling the contributions for the book.

Last but not least, we would like to thank and appreciate our publication partners at Springer, especially Swati Meherishi and Aparajita Singh for helping us bring out this special volume comprising invited and keynotes talks.

About the Conference

ICNETS^2—International Conference on NextGen Electronic Technologies–Silicon to Software is scheduled to be held in VIT University, Chennai from March 23–25, 2017. ICNETS^2 is aimed at providing a forum for engineers, scientists, industry personnel, postdocs and budding researchers to present their research findings and share their knowledge/experience relating to recent advancements and current opinions in the field of Electronics and Communication Engineering. Silicon-Software has been conceived as the theme of this world-class forum covering the imperative aspects of NextGen Electronic Technologies.

It is estimated that more than 400 presenters will attend the conference from diverse background and culture making the event truly international. The conference is comprised of technical sessions (both oral and poster presentations), plenary, keynote addresses and invited talks from reputed authorities around the globe with a vision to provide a networking platform for budding researchers, Ph.D. and postgraduate students. Besides that this global forum is poised to disseminate the knowledge, exchange of skill sets and allied scientific information amongst researchers, developers, engineers, technocrats, professionals, industrialists and government personnel. A joint Private Sector–Academia–Government partnership forum will be convened to discuss the roles and responsibilities of each other as contributing factors towards 2020.

Contents

About the Editors

Dr. S.R.S. Prabaharan is Professor and Dean of School of Electronics Engineering (SENSE), VIT University, Chennai campus since August 2015. In the past he has worked with Manipal International University in Malaysia, the University of Nottingham Malaysia/UK campuses, and CSIR National Laboratories (NAL and CECRI). He is the recipient of Young Scientist Award in the year 1994. He spent a significant time in France and US labs and universities as postdoctoral fellow and technology consultant. He has served as a faculty member in various international universities including University of Malaya, Universiti Teknologi Petronas and Multimedia University, UPMC, Paris, UPJV, Amiens, France and Tohoku University, Japan. He has also served as Adjunct Professor in Malaysia University of Science and Technology and Manipal International University, Malaysia. He has served on various expert committees in Malaysia (Multimedia Development Corporation and Office of the Prime Minister's Department (UNIK) and FP7 European Union committee). He was the technology consultant to United States Army research project under the Southern University, Baton Rouge, USA. He was the recipient of prestigious Erasmus Mundus Visiting Scholar award under the European Union (EU) and held a visiting professorship in France. He is also the recipient of UK-Singapore Visiting Scientist award in 2007 in the University of Sheffield, UK. Recently, he has held a visiting professorship in Tohoku University, Japan. Dr. Prabaharan has received several research grants from DST, DRDO, Young Scientist grant scheme, Malaysian government grants (MOSTI and MOHE) and Motorola Inc., USA. His expertise focuses on supercapacitors for a variety of application including naval missiles (DRDO). His current research interests include memristors and neuristors as well. He has delivered over 52 invited/ keynote talks in international scientific conferences/ symposia. He has edited a few research books and has been the guest editor for international journals as well. He served as consultant to the battery and supercapacitor industries. He has published over 90 high-impact journal papers and over 200 conference papers, and holds international granted patents (Malaysia, USA and Japan). He is known for his notable research in the areas of power sources (Li-Ion batteries and supercapacitors), RF energy harvesting, advanced power solutions for IOT gadgets and memristors.

Nadia Magnenat Thalmann is currently Professor and Director of the Institute for Media Innovation, Nanyang Technological University, Singapore. She is also the Founder and Director of the MIRALab, an interdisciplinary lab in Human Computer Animation, University of Geneva, Switzerland. Her global research area is the modelling and simulation of virtual humans. She is also working on social robots, mixed realities and medical simulation. All over her career, she has received many artistic and scientific Awards, among them the 2012 Humboldt Research Award, and two Doctor Honoris Causa (from University of Hanover in Germany and from the University of Ottawa in Canada). She is Editor-in-Chief of the Journal the Visual Computer (Springer-Verlag) and is a Member of the Swiss Academy of Engineering Sciences.

Dr. V.S. Kanchana Bhaaskaran is Professor in the School of Electronics Engineering and Dean of Academics at VIT University Chennai. She holds an undergraduate degree in Electronics and Communication Engineering from Institution of Engineers (India), Calcutta, an M.S. in Systems and Information from Birla Institute of Technology and Sciences, Pilani, and a Ph.D. from VIT University. She has more than 37 years of industry, research and teaching experience by serving the Department of Employment and Training, Government of Tamil Nadu, IIT Madras, Salem Cooperative Sugar Mills' Polytechnic College, SSN College of Engineering and VIT University Chennai. Her specializations include low power VLSI circuit design, Microprocessor architectures and Linear Integrated Circuits. She has published around 80 papers in international journals and conferences, and has one patent granted and one filed. She has authored two books on linear integrated circuits. She is the Fellow of the Institution of Engineers (India), Fellow of the Institution of Electronics and Telecommunication Engineers, Life Member of the Indian Society for Technical Education, and Member of IEEE.

Chapter 1
Nadine: A Social Robot that Can Localize Objects and Grasp Them in a Human Way

Nadia Magnenat Thalmann, Li Tian and Fupin Yao

Abstract What makes a social humanoid robot behave like a human? It needs to understand and show emotions, has a chat box, a memory and also a decision-making process. However, more than that, it needs to recognize objects and be able to grasp them in a human way. To become an intimate companion, social robots need to behave the same way as real humans in all areas and understand real situations in order they can react properly. In this chapter, we describe our ongoing research on social robotics. It includes the making of articulated hands of Nadine Robot, the recognition of objects and their signification, as well as how to grasp them in a human way. State of the art is presented as well as some early results.

Keywords Robotic hand · 3D printing · Object recognition · Trajectories planning · Natural grasp

1 Introduction

In robotics, the uncanny valley [1] is the hypothesis that human replicas that appear almost but not exactly like real human beings elicit eeriness and revulsion among some observers. Unfortunately, most of the humanoid machines fall into the uncanny valley as they are not exactly like a human (Fig. 1). The uncanny valley is the region of negative emotional response towards robots that seem "almost" human. Movement amplifies the emotional response.

Humanlike Behaviors, such as grasp are vital to solving this problem. Some researchers have done much work on robot grasp, but they usually focus their attention on grasp for industrial robot hands. Most of the hands designed for industrial robots are very different from real human hands. Industrial robot hands

N.M. Thalmann (✉) · L. Tian · F. Yao
Institute for Media Innovation (IMI), Nanyang Technological University (NTU), Singapore, Singapore
e-mail: nadiathalmann@ntu.edu.sg

S.R.S. Prabaharan et al. (eds.), *Frontiers in Electronic Technologies*,
Lecture Notes in Electrical Engineering 433, DOI 10.1007/978-981-10-4235-5_1

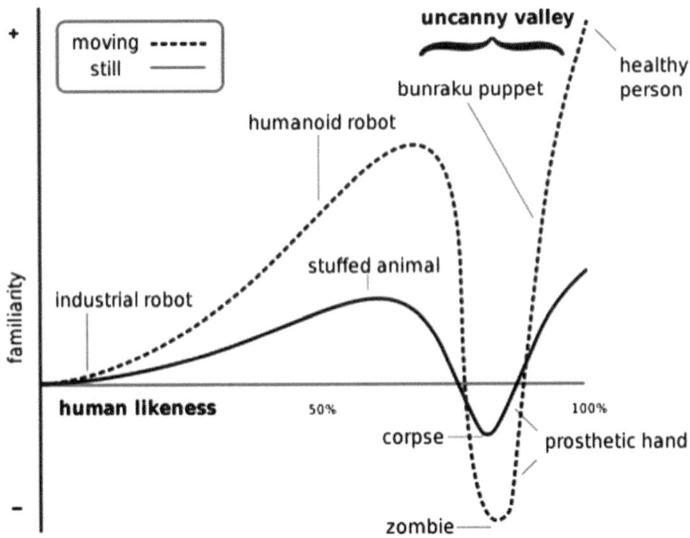

Fig. 1 Uncanny valley [1] Proper hypothesized emotional response of subjects is plotted against anthropomorphism of a robot, following Mori's statements

are usually simpler and may have only two or three fingers, but real humans interact with objects by all fingers most of the time. Therefore, grasping area detection and human-like hands for social robots are more complicated and difficult. In our chapter, we focus on two areas, the making of a proper human-robotic hand and grasping area detection (localization) for pinch grasp. We take pinch grasp as the first example which is most widely used in our common life. Pinch grasp means we hold objects with our thumbs and the opposed fingers.

Grasping is one of the most important abilities for a human being when interacting with objects around them. For humans, it seems quite easy to know what kind of objects there are and grasp them. It is almost immediate. However, for robots, it is quite difficult to grasp something. First, the robots have to know what objects are in front of them. Second, they need to know the location of the objects. Even if the robots know the above information, when interacting with large objects and performing some ordinary tasks, such as picking up a cup full of water, it is difficult to do it well. Robots should neither put their fingers in the water nor cover the top of the cup with their palm. Therefore, the robots should know the location of the proper grasping area and orientation for grasping. In the physical world, the categories and location of objects and grasping area location are unknown. Thus vision is desirable because it provides the information which is more accurate than other information resources. Information based on vision is very necessary. In this chapter, we solve the above mentioned problem with information captured through vision. The entire problem can be divided into two tasks: object recognition, grasping area detection and localization. We will construct a shared neural network for these two tasks. The input is images captured by the Kinect, and the output is the

Fig. 2 Proper grasping area: handle [2]

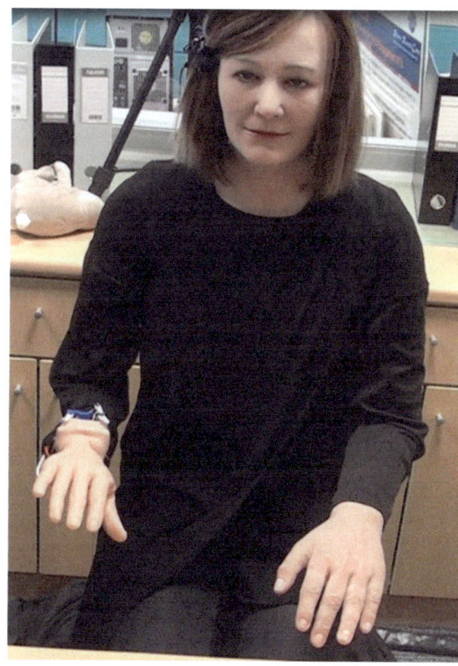

Fig. 3 Nadine robot at IMI, NTU

categories of objects detected, location and orientation of the proper grasping area (Fig. 2).

A robotic hand is a type of mechanical hand, usually programmable, with similar functions as a human hand. The links of such a manipulator are connected by joints allowing either rotational motion (such as in an articulated robot) or translational (linear) displacement. In this chapter, we describe how we have designed a robotic hand for our social robot Nadine (Fig. 3) [3]. When designing the hand, we considered the weight of the robotic hand, the compatibility with Nadine robot and the cost of time and money. Therefore, we made our robotic hand look and work like a real hand as much as possible. After creating the new hand, we tested and modified it to get better grasping force and fingers' trajectories. In the end, the hand was then tested with the Nadine robot to verify our new algorithm of natural grasping approach.

The contributions of this chapter can be summarized as follows:

- A framework which detects object localization and the grasping area in a shared neural network. It is quicker than two single networks and real time.
- A new design of robotic hand for Nadine robot.

The layout of this chapter is as follows. Next section is the related work. Section 3 shows details about object localization and grasping area detection. The design of robotics hand is presented in Sect. 4. Section 5 describes our grasping experiments. Future work and conclusions are in Sect. 6.

2 Related Work

2.1 Humanoid Robot

There are lots of advanced humanoid robots in the world. The Atlas robot [4] developed by Boston Dynamics, the ASIMO [5] robot developed by Honda, the iCub [6] created by the RobotCub Consortium, etc., are all fantastic robots with various kinds of powerful functions. However, they are shelled with plastic or metal which make them look like a robot more than a human. If people stand in front of them, they may be afraid of them. To overcome this feeling, we have chosen a robot with deformable skin which is similar to a real human. Nadine is a realistic female humanoid social robot designed by the Institute for Media Innovation of Nanyang Technological University and manufactured by Kokoro Company in Japan. The robot has strong human-likeness with a natural-looking skin and hair. Kokoro Company has already produced quite a few well-known humanoid robots as the "Actroid" [7] serials. Robot "JiaJia" created by the University of Science and Technology of China is also a good sample of realistic humanoid robots [8]. Hong Kong designer Ricky Ma is also successfully constructing a life-sized robot looking like Hollywood star Scarlett Johansson [9]. They look like human beings, but they are not able to recognize objects and grasp them in a human way. To let Nadine grasp in a human way, we designed two dexterous hands and gave them the ability to grasp like a human being. In this chapter, we show our solution which includes grasp approach using vision-based information. We also first designed our 3D virtual hand to be printed later on (Fig. 4).

Fig. 4 Humanoid robots From *left* to *right*: Atlas robot [4], ASIMO robot [5], iCub robot [6], Actroid robot [7], JiaJia robot [8], Ricky Ma [9]

2.2 *Object Recognition and Localization*

Object localization, or object detection, means finding a predefined object in a set of images. One popular way to detect objects is to use regions with convolutional neural network features (R-CNN) [10] and fast versions [11, 12]. The authors divide object localization problem into two steps: generating object proposals from an image (where they are), and then classifying each proposal into different object categories (object recognition). The major drawback is repeated computation and fast R-CNN that is not fast enough. To improve the speed, we chose first YOLO [13] to eliminate bounding box proposals, but this solution did not perform very well with small objects. We then chose SSD [14] similar to YOLO, but it uses a small convolutional filter to predict object categories and bounding box locations. It allows multiple features maps to perform detection at multiple scales. It increases accuracy, especially for small objects (Figs. 5 and 6).

The second task, grasping area detection using visual information has arose people's interest during the last ten years. Saxena et al. [15] first use visual information to predict grasping region with a probabilistic model (Fig. 7). In [16], the authors try to get multiple grasping points rather than one grasping region (Fig. 8). Nearly all use local hand-designed features to learn good grasping areas. Local features only reflect part of the information, and hand-designed features usually require expert knowledge which makes it difficult to design good features. Therefore, some researchers started to use deep learning to detect grasping area.

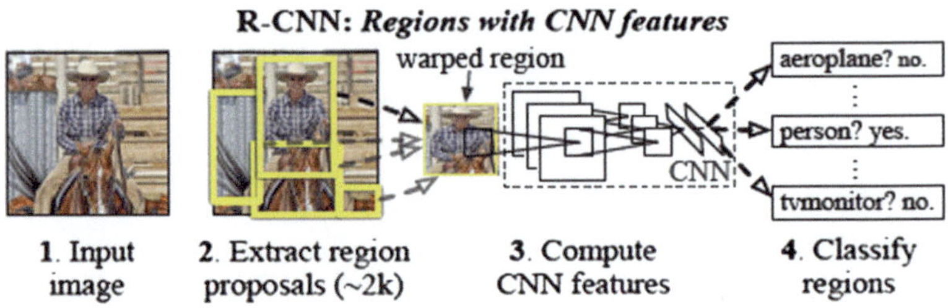

Fig. 5 RCNN workflow [10]

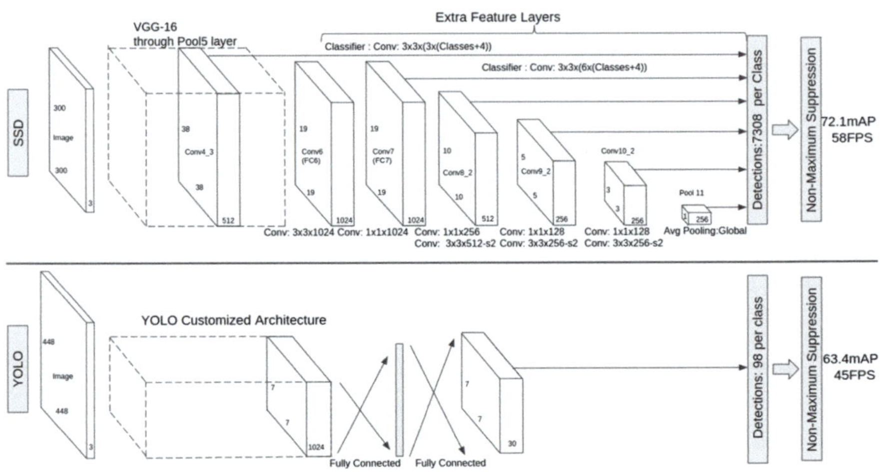

Fig. 6 YOLO [13] and SSD [14] architecture

(a) Martini glass (b) Mug (c) Eraser (d) Book (e) Pencil

Fig. 7 The images (*top row*) with the corresponding labels (shown in *red* in the *bottom row*) of the object classes used for training [15]

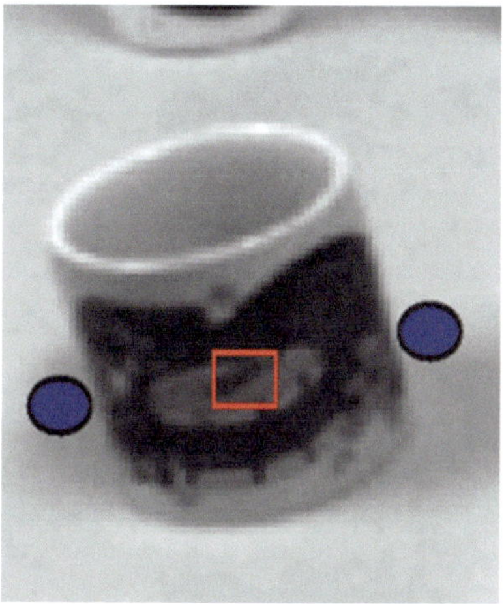

Fig. 8 Finger contact points (*blue circles*) [16]

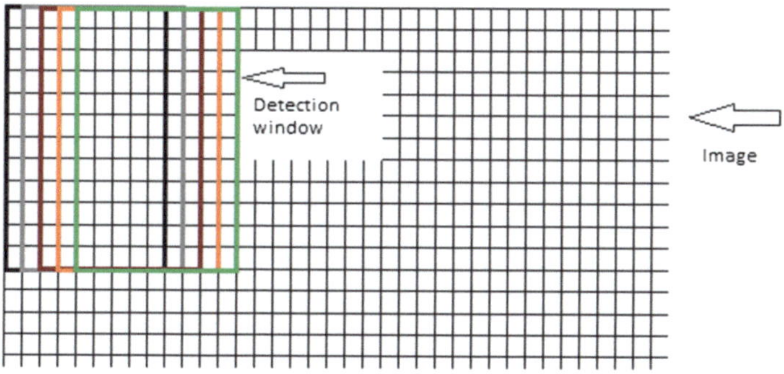

Fig. 9 Sliding window [21]

Lenz et al. [17] use convolutional networks as a classifier for grasping area detection in sliding windows. The sliding window means getting image patches when a window slides from left to right and from top to bottom and then independently classifying all image patches as being good grasping areas or not (Fig. 9). Detecting good grasping area in sliding windows is time-consuming because there are lots of image patches in an image. The authors trained their neural network model on Cornell Grasp Dataset [18] which is the most widely used datasets. Redmon et al. [19] also use the same dataset and deep convolutional network for this task. The solution is real time but not accurate, especially for small objects since it divides one image into several grids and predicts one grasp per grid. Guo et al. [20] uses similar neural network architecture which could also detect objects and grasping areas and create a new dataset. However, their solution only works when objects are put on a horizontal plane. Their dataset only contains fruits so their solution may not detect other objects in our daily life. We build a shared neural network which is similar. It can perform object recognition and grasping area

Table 1 Approaches comparison for grasping area detection

Approach	Contribution	Limitation
Robotic grasping of novel objects using vision [15]	Can get grasping points	Grasping points are not enough, more information needed, such as orientation
Deep learning for detecting robotic grasps [17]	Can get grasping parts	No object recognition
Real-time grasp detection using convolutional neural networks [19]	Combine object recognition and grasping area detection	Perform not well for small objects
Object discovery and grasp detection with a shared convolutional neural network [20]	Combine object recognition and grasping area detection	Could only grasp objects on horizontal planes
		Could only work for fruits grasp
Our approach	1. Combine object recognition and grasping area detection in a shared CNN	
	2. Perform well even for small objects	
	3. Could work for any graspable objects	
	4. Could work for objects placed everywhere	

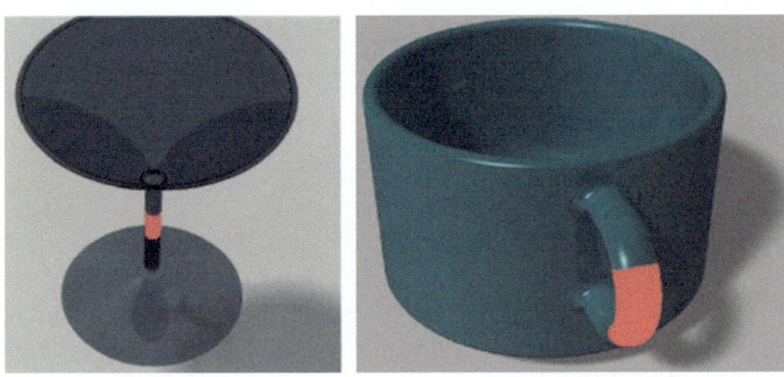

Fig. 10 Grasping region in Saxena et al. [15]

detection simultaneously and works well with any graspable objects placed on any plane, even for small objects. A detailed comparison is shown in Table 1 (Figs. 10, 11 and 12).

2.3 *Humanoid Robotic Hands*

For the design of robotic hands, most existing robotic arms or hands can only perform pre-programed actions, especially for industrial robotic arms. Only a few

Fig. 11 Grasping *rectangle* in Lenz et al. [17]

Fig. 12 Cornell Grasp Dataset [18]

ones can do autonomous grasping, but the action is inflexible regarding handling the complex objects. Some of them show wonderful high speed and precise movement. The KUKA robot [22] can beat famous table tennis player Timo Boll (Fig. 13). Motoman-MH24 [23] is a Sword-wielding robot which beat Japanese master samurai (Fig. 13). However, these robotic hands are very different from real human hands regarding shape and behavior. Bebionic V3, Dextrus and Tact (Fig. 14) are 3 leading anthropomorphic hands. In this chapter, Inmoov hand, i-LIMB hand (Fig. 14), are also reviewed to compare with Nadine's hand as well in Sect. 4.

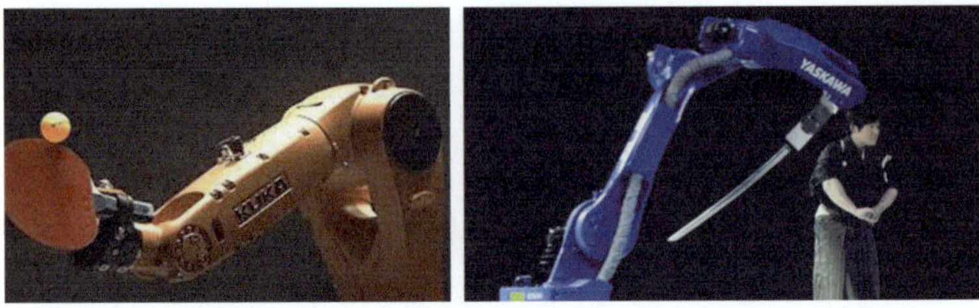

Fig. 13 The KUKA robot and Motoman-MH24 [22, 23]

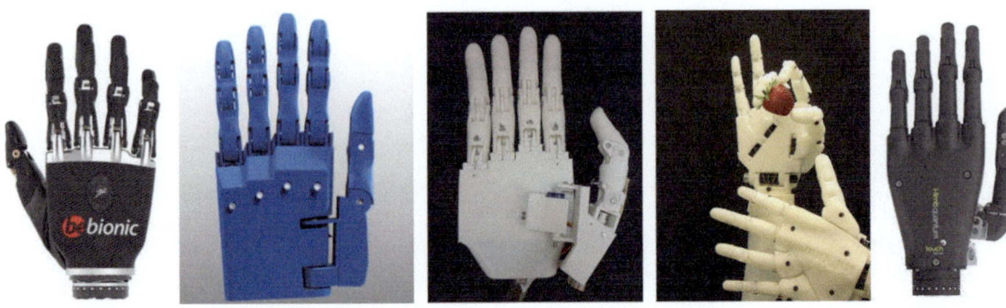

Fig. 14 (From *left* to *right*) Bebionic, Dextrus, Tact, Inmoov, and i-Limb hand [24–26]

3 Object Localization and Grasping Area Detection

3.1 Grasp Representation and Accuracy Metric

Using vision-based information, we want to detect all objects and proper grasping area. The vision nsdrf information includes images and points cloud data. We use color images and depth images from Kinect because it is accessible and easy to process. In our chapter, we focus on pinch grasp as pinch is most widely used by our human being in common life, and it is easier to study compared to other complicated grasp types which may involve multiple fingers. For pinch grasp, five-dimension representation (Fig. 15) was proposed by Lenz et al. [17].

This representation can be expressed as {x, y, θ, w, h}. x, y here are the center coordinates of the grasping rectangle. θ is the orientation angle of this rectangle. W and h are width and height of the rectangle. Figure 15 is an example of this grasp representation.

To evaluate the accuracy of our predicted grasping area, we adopted a rectangle metric used by Lenz et al. [17] and Redmon et al. [19]. The metric classifies a grasp rectangle as correct and valid if:

- The angle between the ground truth rectangle and predicted rectangle is less than 30°

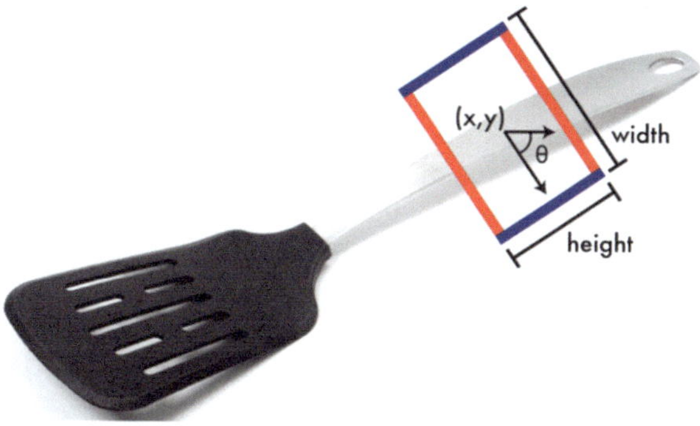

Fig. 15 Pin representation [19]. The *blue edges* are parallel to pinching fingers and width is the open width of the two fingers. The *blue lines* tell the size and orientation of the two pinching fingers. The *red lines* show the distance between thumbs and the opposed fingers

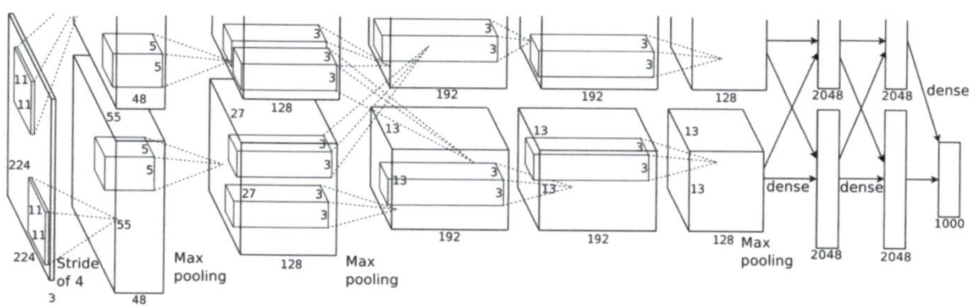

Fig. 16 AlexNet network architecture [27]

- The Jaccard index $\frac{|A \cap B|}{|A \cup B|}$ is greater than 0.25. A and B are ground truth rectangle and predicted rectangle
- The metric requires that a correct grasp rectangle should have large common areas with ground truth rectangle and their orientation rectangle should be small.

3.2 Model Architecture

Our convolutional neural network is a direct solution from color and depth images to object categories and grasping areas. We do not need object detection, and we predict proper grasping areas and object categories directly. Our network outputs a grasping rectangle which represents the proper grasping area and categories of objects on which rectangles are located. It is a combined convolutional neural network which could finish two tasks: object recognition and grasping area detection. This idea is derived from [19]. The combination decreases the training and testing time of the convolutional neural network and makes it real time without sacrificing accuracy.

The first five layers derived from AlexNet extract common features for object classification and grasping area regression. AlexNet (Fig. 16), one well known and widely used convolutional neural network, is first proposed in [27]. AlexNet is designed for object recognition and gets very good scores in the ImageNet competition. We adopted the AlexNet because we could easy fine-tune our convolutional neural network and fine-tuning could greatly improve the accuracy of results and eliminate overfitting. The architecture can be found in Fig. 16. They are five convolutional layers and two connected hidden layers in AlexNet. We copy the first convolutional layers and add two connected hidden layers with 2048 neurons each. At the end of the network, there is a softmax loss layer [28] which outputs object categories and a regressor, a Euclidean loss layer which outputs predicted grasping area.

3.3 Data Preprocessing and Training

Only a few datasets are suitable for our work. We chose Cornell Grasping Dataset [18]. There are 885 images of 240 different objects. For each image, there are one color and one depth image and 5 or more labeled grasping rectangle. It is originally designed for grippers, but it is also suitable for our use. We divide the images into about 16 categories, such as "bottle," "shoes" and so on.

To make full use of depth information, we substituted the blue channel with the depth channel. Then, we took a center crop of 340 * 340 for each image and did data augmentation for them. We translated them with a random number (between 0 and 50) of pixels in x and y direction and rotated them up to 360°.

We used our convolutional neural network showed in Fig. 3. We fine-tuned our model with weight file from AlexNet. As the first five layers in our network are the same as that in AlexNet, so we could initialize the first five layers with AlexNet weights. Then we will freeze the first five layers and train the network with the Cornell grasping dataset.

3.4 Software Development

We first developed our software using Theano [29], a deep learning python package. We implemented full network architecture from the beginning, preprocessing our data, defining different layers, training and testing the network and visualize our result. However, the network performance was too slow for training, which took about one day for training once. We then implemented our whole program with Caffe [30], a well-known deep learning platform. It provides lots of pre-trained neural network model which could be used for fine-tuning, and it is also much faster than Theano.

4 Nadine Hand Design

In this section, we identify the biomechanical features that affect the functionality of human hand from the aspects of the bones and the joints. Furthermore, we introduce the model of Nadine hand, which drastically reduces the degrees of freedom (DOF) of the real human hand.

4.1 Human Hand Features

Figure 17 displays the skeleton of a real human hand which is made of 27 pieces of bones. The five fingers are constituted by eight small wrist bones, five metacarpal bones (in palm) and 14 finger bones. The bones are connected by five distal interphalangeal (DIP) joints, four proximal interphalangeal (PIP) joints and five metacarpophalangeal (MCP) joints in hand. The thumb is a special case which has no DIP or PIP but an Interphalangeal (IP) joint, a trapezio-metacarpal (TM) joint and an MCP joint. Thus, the real human hand has 15 movable joints in total and each finger have 3 of them.

A human hand totally has 27 DOF in total. There are 5 DOF in the thumb, 4 DOF in each other finger and 6 DOF in the wrist [31]. The 6 DOF in the wrist have the property of "global motion" which refers that all the palm and fingers will move at the same time, while the other 21 DOF in hand has the property of "local motion." In most prior studies of the motion or the design of robotic hands, only the "local motion" of the hand is considered.

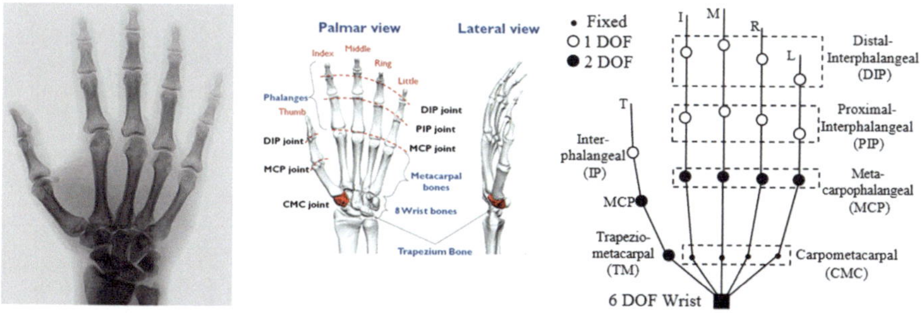

Fig. 17 Bones in human hand and DOF in hand [31]

4.2 Human Hand Constraints

Based on our observation, we find that the human hand and fingers have the following constraints:

- Fingers cannot bend backward without external force.
- The movement of the pinky finger will affect the ring finger or even the middle finger. Also, all the other four fingers are affected if the middle finger is bent.
- The DIP joint of each finger cannot be moved alone.

A formal representation of the constraints of human hands is proposed by Lin et al. [32], in which the constraints are divided into three types:

Type I has static constraints that limit the finger motion as a result of the hand anatomy. For example, the DIP joints cannot be moved more than 90°.

Type II has dynamic constraints that limit the joints in motion. For instance, when we bend a DIP joint, the corresponding PIP joint must be bent too.

Type III has the constraints that exist in certain human hand gestures. Sometimes they are not hard constraints, but most of the human beings follow it. For example, to make a fist, most people bend all the fingers together instead of bending fingers one by one.

4.3 Reduction of the Hand DOF

In this section, we propose several heuristics to reduce the DOF of the hand model. Firstly, we can reduce 5 DOF of the human hand by ignoring the subtle abduction/adduction motion of the fingers' MCP joints. So we reduce the DOF of the hand's local motion to 16. Moreover, only the thumb's TM joint has 2 DOF. It is difficult to simulate 2 DOF on one joint due to mechanical limitations. In Nadine's hand, a new joint is added to represent the TM joint of the thumb and simulate the abduction/adduction moving of the thumb. The final DOF is 16 and showed in Fig. 18.

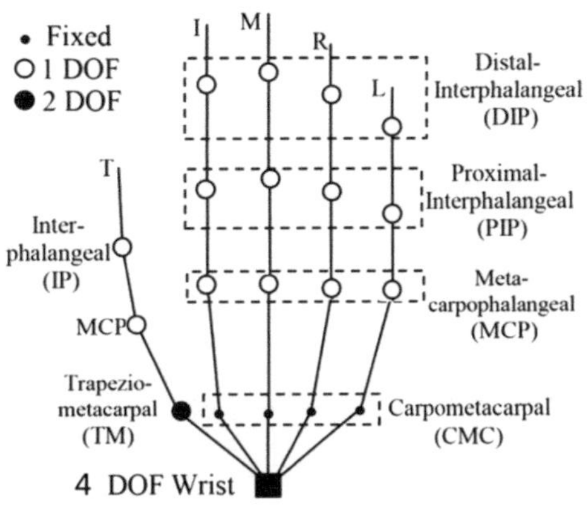

Fig. 18 Mechanical DOF in Nadine's hand

For the wrist part, with the help of Nadine robot's arm motion, Nadine can move hand up/down, side to side, forward/backward which contribute 3 DOF. However, as the design limitation of Nadine robot, it has only one actuator to simulate flexion/extension between the carpals and radius. So the abduction/adduction and supination/pronation motion from the wrist bone are omitted, and there are 4 DOF to represent the global motion of the hand. This omission will not affect the hand global motion much as arm will assist the hand to make the global motion. Together with the local motion, there are totally 20 DOF for Nadine's hand.

4.4 *Simulate the Local Motion of the Human Hand/Finger*

There are many ways to simulate the local motion of the hand. The difficult part is how to reproduce the DOF and the motion angles of each finger. It is also a tradeoff between performance, ability, and efficiency. One simple way is to use 16 actuators to control 16 DOF local motion. It will get direct control of each DOF separately. However, so far none of the commercial anthropomorphic hand chooses this way as the cost and design complexity are extremely high. The thing goes the same for Nadine's hand. So the first problem is to simulate the local motion with a minimal number of actuators (Fig. 19).

Nadine's hand has 6 actuators to control 16 joints and 6 DOF. The actuate way of Nadine's hand is similar like Dextrus [26] or Inmoov hand [24]. For the thumb part, the Inmoov hand's thumb has three joints but only have 1 DOF in flexion/extension. The Dextrus hand's thumb has 2 DOF in flexion/extension and abduction/adduction, but its flexion/extension has only two joints (IP and MCP). Nadine's hand combines the advantage of both hands. Its thumb has three joints and all three joints controlled by 1 DOF in flexion/extension. It has one more joint which split from the TM joint of the thumb so that it can make the

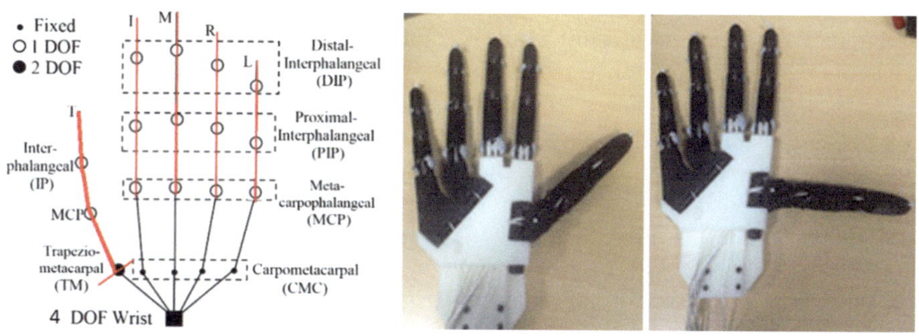

Fig. 19 Actuate DOF in Nadine's hand

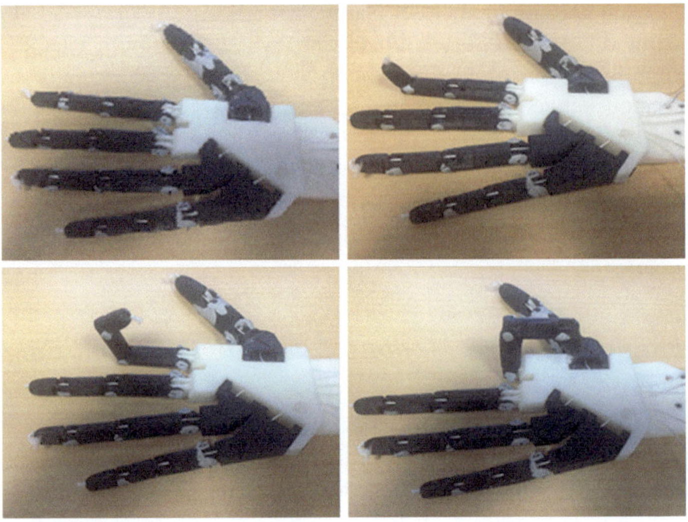

Fig. 20 Actuated fingers of Nadine's hand

abduction/adduction move. These joints can work under the artificial skin without any abnormal in the appearance of the hand.

In Fig. 20, we show the internal design of Nadine's hand and how it works. For each finger, it can be assembled by three pieces of 3D printable parts. The joint part can be linked together by a metal screw or a piece of solid wire like iron wire. A thread like a fish line or cotton wire goes through the inner part of the whole finger to control the movement. When pulling the control thread, the finger will move starting from the joint with minimal resistance force. Different weights in these three 3D printed parts represent the distal, middle and proximal phalanges. The distal part should move first as it is the lightest part and it has the minimal resistance force. Then if the pulling continues, the distal part will continue to move until it reaches the maximum moving point. After that, the middle phalange should start to move and stop at the limit angle. The proximal phalange should move last as it has the maximum resistance force. Nadine's hand only has one thread for each finger for flexion. As Nadine's hand has artificial skin, it provides the elastic force to make the finger extension after the release of the thread. It makes the design of the actuator simpler. Table 2 shows the Nadine hand's joint moving angles compare to other robotic hands.

As Nadine robot uses the air motor for its actuator, the weight of the new hand should be as light as possible to avoid making a heavy load to the joint of the wrist. It is the reason to choose plastic rather than metal. The technology of 3D printing gives an easy way to design and try errors. As most of the 3D printed hand has the same size as a human, it is impossible to put inside the artificial hand skin like

Table 2 Hand joint moving angle

Hand	Metacarpophalangeal joints (°)	Proximal interphalangeal joints (°)	Distal interphalangeal joints (°)	Thumb flexion (°)	Thumb circumduction
Nadine hand	0–90	0–110	0–90	0–90	0–90
Tack [25]	0–90	23–90	20	0–90	0–105
Dextrus [26]	0–90	0–90	0–90	0–90	0–120
I-Limb pulse [26]	0–90	0–90	20	0–60	0–95
Bebionic V2 [26]	0–90	0–90	20	–	0–68

"InMoov hand" [24] and "Tack hand" [25]. Based on the study of the state-of-art 3D-printed hand, the new created Nadine's hand is smaller and slimmer and does fit the artificial hand skin. With the help of 3D modeling software like 3DS Max or AutoCAD, the size of the hand can be easily changed to different scales, as well as the length of each bone. In the future, we could design software to automatically generate the 3D printed part of customized size hand by giving some basic size of the finger like each finger's and palm's dimensions.

For the weight, Nadine's hand is 200 g and much lighter than any of the existing hand. The three main reasons are first the 3D printed parts are not heavy as the metal parts in the commercial hand. The second reason is the servo motor of Nadine's hand used for the actuator (HITEC HS-5070MH) is only 12.7 g each. The third one is that Nadine's hand can use the external power. There is then no need to have a battery inside the hand.

5 Grasping Experiments

In this section, the characters of human grasping are listed and applied to virtual Nadine. After that, the grasp experiment is tested on Nadine robot with new robotic hand.

5.1 *Human Grasps Motion Study*

The human hand can move up to 50 m/s. However, most of daily motion's speed is within the range of 0.5–5.0 m/s. It will be abnormal if the hand move too slow or too fast without a reasonable purpose. Nadine robot's forearm can at most move roughly 3 m/s under current air pressure. Moreover, it is easy to slow down the moving speed of Nadine by software method. We can also try to move Nadine's forearm and arm simultaneously to get the faster speed for moving the hand.

When humans want to move the hand to a certain position, he/she will use the joint of the shoulder, elbow or maybe wrist at the same time. Furthermore, the separate joints' motor will take almost the same time. For example, if a person wants to move his/her hand in a position which requires moving elbow up to 120° and shoulder up to 10°, he/she will not move them one by one, nor move them at the same speed. In that case, the shoulder will finish the movement much earlier than the elbow. The elbow and shoulder will cooperate with each other and try to have the same start and end time. Nadine's motion parameters are provided for each piece and each piece is performed in sequence.

When humans try to grasp something, their sight is most likely focused all the time on the target during the action. Nadine robot has its vision system. The first step of grasping action is to turn the head and eyes to locate the target. At this stage, the position of the target is fixed on the table. The first motion of Nadine is to low down the head and put the sight focus on the target and continue with the hand's motion.

Muscles and tendons support human to complete the grasp action. They are linked together, and they also constrain our motions. So a human cannot reach or hold some position too long as it will stretch or curl muscles or tendons too much. It is also needed to avoid the motion beyond the human anatomy limitation.

Humans usually show different face emotions when doing some action. As Nadine robot has seven movable points in the face, it adds to the human grasp motion to simulate some emotions.

5.2 *Motion Design on Virtual Nadine*

The Nadine robot has 27 air motors, and each of them can be set to different values to achieve various postures. Up to 30 posture frames can be set within one second. Then smooth actions can be produced if these postures are playing in sequence. To make the action look more human like, moving speed, motion continuity, motion simultaneity, sight focus and coordinate body movement are considered. The first action indicates the default position of Nadine robot with two hands in front of the body. The second action is for "eye on the target." Nadine robot will head down and look at the table. The third action is for "approach to the target," Nadine robot will forward her right hand higher and forward to the target object. The fourth action is for "touch the target," and Nadine robot will put her hands to the suitable position before grasp. Additional head, waist, and left arm/shoulder motions are

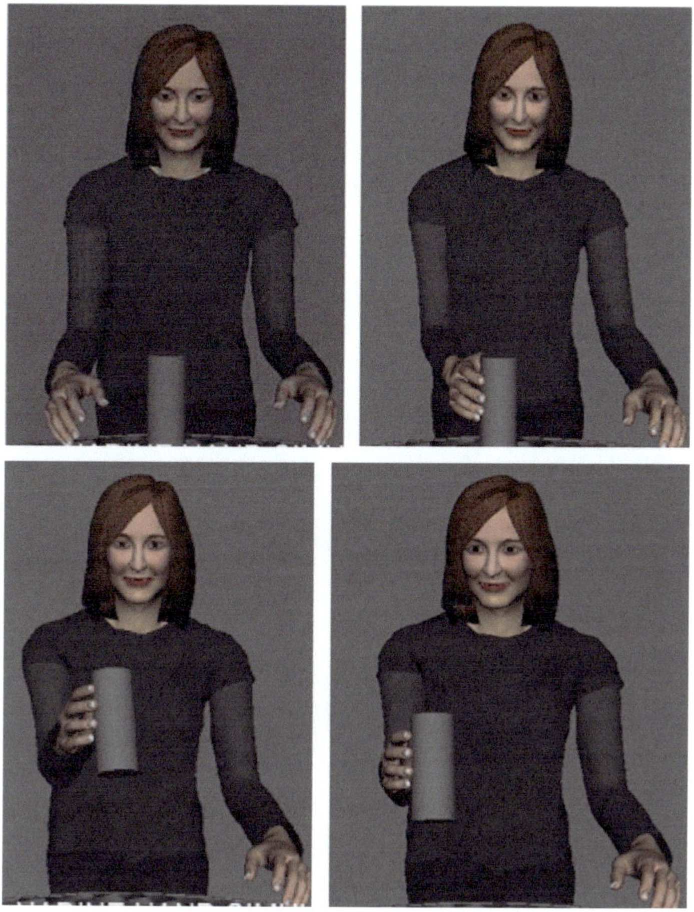

Fig. 21 Four intermediaries' actions of Virtual Nadine

added to third action to make the whole process more coordinated and human like. Some facial emotions are also added.

Four intermediaries' actions are showed in Fig. 21.

5.3 Grasp Experiments of Nadine

Nadine's hand is tested inside Nadine's artificial hand skin. A 3 × 3 × 5 CM sponge toy is used as grasp target. From Fig. 23, Nadine can move the hand toward the object, grasp it and move it to a new predefined location. These actions are similar to the virtual Nadine showed in Fig. 21. Four intermediaries' actions are showed in Fig. 22 for the real Nadine robot.

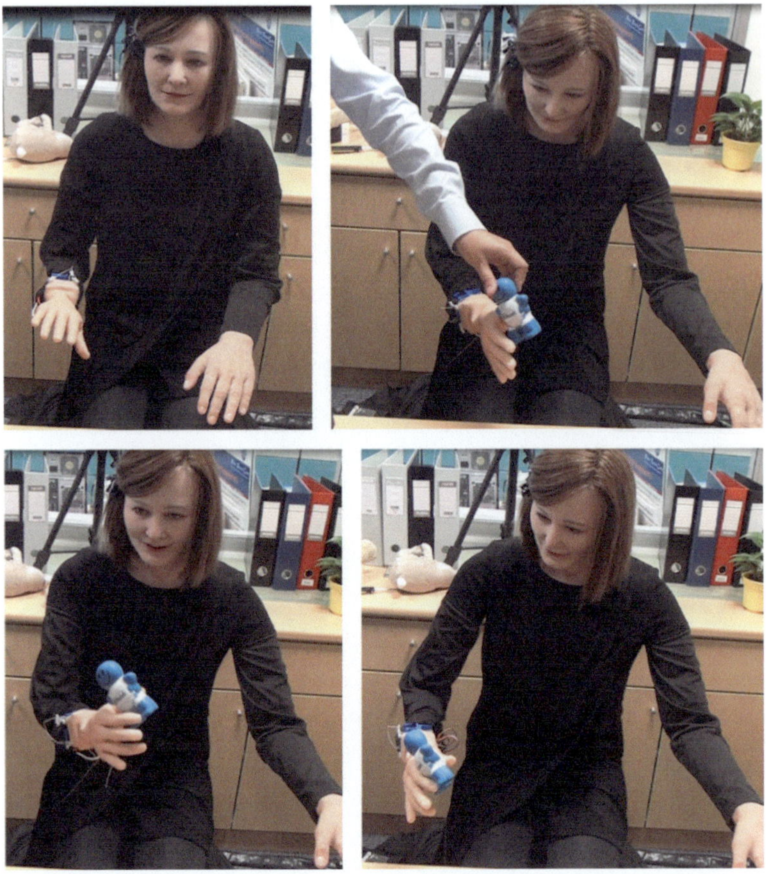

Fig. 22 Four intermediaries' actions of Nadine robot

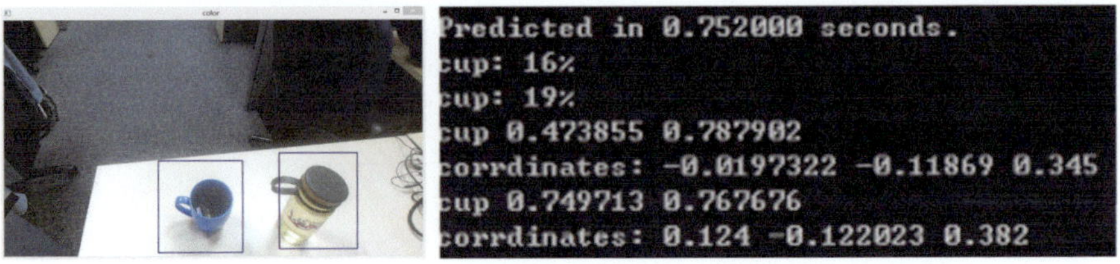

Fig. 23 Object detection's result

5.4 *Grasp Experiments with Visual Information*

Our experiments about grasping area detection are still on development. We are using visual information to guide our grasp. We first set up a Kinect v2 on a desktop in front of Nadine. After our neural network predicts the object categories and 2d position, we will convert the 2d position to 3d position and orientation according to surface normal vectors obtained from depth images. The robot then will reach out her hands to the position and grasp the object.

The following is the object detection's results. It is real time and can achieve high accuracy.

After 15000 iterations for training our model and testing it on 300 images, we find that it can predict the location of the object, but the orientation and location of the grasping area are not accurate enough. These experiments are still under development.

6 Conclusion and Future Work

We presented our ongoing research on human-like grasp for social robots. It includes our state-of-the-art methods for the design of a robotic hand, the object recognition and grasping area detection. With a large number of experiments, we successfully validated our methods. The robot Nadine can achieve the grasping with a high accuracy as well as behave human-like when grasping. There is still a lot of work to do in this area. For example, how to make the grasp motion more human like rather than reaching directly to the destination with a constant speed? It is still a problem to solve.

In summary, our future work will mainly focus on the following areas:

1. Improve the robotic hand. Our current hand is not powerful and precise enough for wide and long-time use in common life.
2. Integrate visual information with a robotic hand. The robot should grasp any graspable objects once they come into Nadine's field of view.
3. Improve the way Nadine grasps. There are still a lot to do to make the grasping process completely humanlike.

Acknowledgements This research is supported by the BeingTogether Centre, a collaboration between Nanyang Technological University (NTU) Singapore and University of North Carolina (UNC) at Chapel Hill. The BeingTogether Centre is supported by the National Research Foundation, Prime Minister's Office, Singapore under its International Research Centres in Singapore Funding Initiative.

References

1. Uncanny valley, in Wikipedia. https://en.wikipedia.org/wiki/Uncanny_valley. Accessed 28 Nov 2016
2. https://funwithluka.files.wordpress.com/2015/03/img2885.jpg. Accessed 13 Nov 2016

3. Nadine social robot, in Wikipedia. https://en.wikipedia.org/wiki/Nadine_Social_Robot. Accessed 28 Nov 2016
4. Atlas (robot), in Wikipedia. https://en.wikipedia.org/wiki/Atlas_(robot). Accessed 30 Nov 2016
5. Y. Sakagami, R. Watanabe, C. Aoyama et al., The intelligent ASIMO: system overview and integration, in *IEEE/RSJ International Conference on Intelligent Robots and Systems, 2002*, vol. 3, (IEEE, 2002), pp. 2478-2483
6. G. Metta, G. Sandini, D. Vernon, L. Natale, F. Nori, The iCub humanoid robot: an open platform for research in embodied cognition, in *PerMIS'08 Proceedings of the 8th Workshop on Performance Metrics for Intelligent Systems,* pp. 50–56
7. Actroid, in Wikipedia. https://en.wikipedia.org/wiki/Actroid. Accessed 28 Nov 2016
8. T. Times, Meet Jiajia, china's new interactive robot. http://www.techtimes.com/articles/150827/20160416/meet-jiajia-chinas-new-interactive-robot.htm. Accessed 30 Nov 2016
9. Home, http://www.rickyma.hk/. Accessed 30 Nov 2016
10. R. Girshick, J. Donahue, T. Darrell, J. Malik, Rich feature hierarchies for accurate object detection and semantic segmentation, in *CVPR* (2014)
11. R. Girshick, Fast r-cnn, in *Proceedings of the IEEE International Conference on Computer Vision* (2015), pp. 1440–1448
12. S. Ren, K. He, R. Girshick, J. Sun, Faster r-cnn: towards real-time object detection with region proposal networks, in *Advances in Neural Information Processing Systems* (2015), pp. 91–99
13. J. Redmon, S. Divvala, R. Girshick, A. Farhadi, You only look once: unified, real-time object detection, in *CVPR* (2016)
14. W. Liu, D. Anguelov, D. Erhan et al., SSD: single shot multibox detector (2015). arXiv:1512.02325
15. A. Saxena, J. Driemeyer, A.Y. Ng, Robotic grasping of novel objects using vision. Int. J. Rob. Res. **27**(2), 157–173 (2008)
16. Q.V. Le, D. Kamm, A.F. Kara, A.Y. Ng, Learning to grasp objects with multiple contact points, in *IEEE International Conference on Robotics and Automation (ICRA)* (2010), pp. 5062–5069
17. Ian Lenz, Honglak Lee, Ashutosh Saxena, Deep learning for detecting robotic grasps. Int. J. Rob. Res. **34**(4–5), 705–724 (2015)
18. Y. Jiang, S. Moseson, A. Saxena, Efficient grasping from rgbd images: learning using a new rectangle representation, in *IEEE International Conference on Robotics and Automation (ICRA)* (2011), pp. 3304–3311
19. J. Redmon, A. Angelova, Real-time grasp detection using convolutional neural networks, in *IEEE International Conference on Robotics and Automation (ICRA)* (2015), pp. 1316–1322
20. D. Guo, T. Kong, F. Sun et al., Object discovery and grasp detection with a shared convolutional neural network, in *2016 IEEE International Conference on Robotics and Automation (ICRA)*. (IEEE, 2016), pp. 2038–2043
21. Adioshun, Results matching, https://adioshun.gitbooks.io/learning-opencv-3-computer-vision-with-python/content/learning_python_opencv_ch07.html. Accessed 21 Dec 2016
22. G. Schreiber, A. Stemmer, R. Bischoff, The fast research interface for the kuka lightweight robot, in *IEEE Workshop on Innovative Robot Control Architectures for Demanding (Research) Applications How to Modify and Enhance Commercial Controllers (ICRA 2010)* (2010), pp. 15–21
23. S. Rueckhaus, MH24. http://www.motoman.com/industrial-robots/mh24 Accessed 28 Nov 2016
24. Inmoov project, http://inmoov.fr/. Accessed 1 Dec 2016
25. P. Slade et al., Tack: design an performance of an open-source, affordable, myoelectric prosthetic hand, in *2015 IEEE ICRA*
26. J.T. Belter et al., Mechanical design and performance specifications of anthropomorphic prosthetic hands: a review. J. Rehabil. Res. Dev. **50**(5) 2013

27. A. Krizhevsky, I. Sutskever, G.E. Hinton, Imagenet classification with deep convolutional neural networks, in *Advances in Neural Information Processing Systems* (2012), pp. 1097–1105
28. Layer catalogue, http://caffe.berkeleyvision.org/tutorial/layers.html. Accessed 28 Dec 2016
29. Theano 0.8.2 documentation (2008), http://deeplearning.net/software/theano/. Accessed 21 Dec 2016
30. Deep learning framework (no date), http://caffe.berkeleyvision.org/. Accessed 21 Dec 2016
31. A.M.R Agur, M.J. Lee, *Grant's Atlas of Anatomy 10th ed* (1999)
32. J. Lin, Y. Wu, T.S. Huang, Modeling the constraints of human hand motion, in *Workshop on Human Motion, 2000*. Proceedings. (IEEE, 2000), pp. 121–126

Chapter 2
Data-Driven Analysis of Collections of Big Datasets by the Bi-CoPaM Method Yields Field-Specific Novel Insights

Basel Abu-Jamous, Chao Liu, David J. Roberts, Elvira Brattico and Asoke K. Nandi

Abstract Massive amounts of data have recently been, and are increasingly being, generated from various fields, such as bioinformatics, neuroscience and social networks. Many of these big datasets were generated to answer specific research questions, and were analysed accordingly. However, the scope of information contained in these datasets can usually answer much broader questions than what was originally intended. Moreover, many existing big datasets are related to each other but have different detailed specifications, and the mutual information that can be extracted from them collectively has been not commonly considered. To bridge this gap between the fast pace of data generation and the slower pace of data analysis, and to exploit the massive amounts of existing data, we suggest employing

---

Professor Nandi is a Distinguished Visiting Professor at Tongji University, Shanghai, China. This work was partly supported by the National Science Foundation of China grant number 61520106006 and the National Science Foundation of Shanghai grant number 16JC1401300.

B. Abu-Jamous · C. Liu · A.K. Nandi (✉)
Department of Electronic and Computer Engineering,
Brunel University London, London, UK
e-mail: asoke.nandi@brunel.ac.uk

C. Liu
e-mail: chao.liu@brunel.ac.uk

B. Abu-Jamous
Department of Plant Sciences, University of Oxford,
South Parks Road, Oxford OX1 3RB, UK
e-mail: basel.abujamous@plants.ox.ac.uk

D.J. Roberts
Radcliffe Department of Medicine, University of Oxford,
Headington, Oxford OX3 9DU, UK
e-mail: david.roberts@ndcls.ox.ac.uk

D.J. Roberts
National Health Service Blood and Transplant, John Radcliffe Hospital,
Headington, Oxford OX3 9BQ, UK

S.R.S. Prabaharan et al. (eds.), *Frontiers in Electronic Technologies*,
Lecture Notes in Electrical Engineering 433, DOI 10.1007/978-981-10-4235-5_2

data-driven explorations to analyse collections of related big datasets. This approach aims at extracting field-specific novel findings which can be revealed from the data without being driven by specific questions or hypotheses. To realise this paradigm, we introduced the *binarisation of consensus partition matrices (Bi-CoPaM)* method, with the ability of analysing collections of heterogeneous big datasets to identify clusters of consistently correlated objects. We demonstrate the power of data-driven explorations by applying the Bi-CoPaM to two collections of big datasets from two distinct fields, namely bioinformatics and neuroscience. In the first application, the collective analysis of forty yeast gene expression datasets identified a novel cluster of genes and some new biological hypotheses regarding their function and regulation. In the other application, the analysis of 1,856 big fMRI datasets identified three functionally connected neural networks related to visual, reward and auditory systems during affective processing. These experiments reveal the broad applicability of this paradigm to various fields, and thus encourage exploring the large amounts of partially exploited existing datasets, preferably as collections of related datasets, with a similar approach.

Keywords Data-driven analysis • Tunable consensus clustering • Bi-CoPaM • Heterogeneous datasets • Gene expression • fMRI

1 Introduction

Advances in data collection and storage have generated huge amounts of big data in various research fields. For instance, tens of thousands of gene expression datasets have been generated during the last couple of decades. Each one of these datasets typically measures the expression of thousands to tens of thousands of genes over a number of samples which ranges from two to few hundreds [1]. In neuroscience, a typical magnetic resonance imaging (MRI) scan, covering the whole human brain, consists of millions of voxels, i.e. unitary three-dimensional sections, representing

E. Brattico
Center for Music in the Brain (MIB), Department of Clinical Medicine,
Aarhus University, Aarhus, Denmark
e-mail: elvira.brattico@clin.au.dk

E. Brattico
The Royal Academy of Music Aarhus/Aalborg, Aarhus, Denmark

E. Brattico
AMI Centre, School of Science, Aalto University, Espoo, Finland

A.K. Nandi
The Key Laboratory of Embedded Systems and Service Computing,
College of Electronic and Information Engineering, Tongji University,
Shanghai, People's Republic of China

different parts of the brain. Moreover, the functional MRI (fMRI) technology yields even larger amounts of data as it usually generates hundreds of 3D images along the dimension of time for the same imaging object, i.e. the brain, despite the fact that its individual images have relatively lower resolution than the MRI's. Many other fields yield high-throughput data including social networks [43], and economics [20], to name a few.

Information engineering techniques, such as machine learning and statistical methods, aim at identifying meaningful findings from data by computational analysis. Such techniques are required to analyse any single big dataset where it is not feasible to interpret it manually. This has been done very frequently in the literature when a research group produces a big dataset, analyses it using information engineering techniques, identifies novel conclusions, and publishes the conclusions alongside the dataset for the use of the community. Despite the numerous benefits of this approach, there are a few issues that have arisen over the years and decades. One issue is that many of these big datasets were generated by experiments designed to answer specific research questions or to test specific research hypotheses. Such datasets are commonly analysed in a relatively focused manner while being driven by these questions and hypotheses. Although the big dataset may well answer the question in hand, it usually has a much larger potential than that. Such partially explored datasets have been rapidly accumulating in public and private data repositories.

Moreover, answering a given research question in many cases requires the generation of a big dataset designed differently from any existing dataset. Nonetheless, it may well be related to some existing datasets which do not fully suffice to answer that question. Because of that, many existing datasets are found to belong to the same context but with differences in some details such as the conditions and the resolution of their samples, the used technology, the number of replicates and others. These collections of related datasets definitely carry mutual information which may not be inferred except by analysing them collectively. However, differences amongst such related datasets make them, in many cases, uncombinable in a straightforward concatenating manner. Information engineering techniques that are able to analyse multiple heterogeneous, yet related, datasets collectively have become a necessity to unearth the valuable mutual information buried in these groups of datasets.

In other instances, although a very big single dataset is generated, it semantically consists of many sub-datasets. In other words, it can be split into many big datasets where each one these datasets is semantically an independent dataset. Nonetheless, such big sub-datasets are indeed related to each other in some sense. This is therefore another example where methods that can analyse multiple datasets collectively are required.

In summary, it is evident that what we have learnt from the huge amounts of available big data does not match the potential that such data possess, as individual datasets or as collections of related datasets. We believe that revisiting existing data with novel and able sophisticated computational methods that have exploratory capabilities is crucial to try to bridge the gap between the rapid pace of big data generation and the slower pace of data analysis and information inference. Such efforts should be in collaborations between information engineers and the researchers related to the specific fields of the datasets. Importantly, such efforts should ideally be exploratory and data-driven instead of guided hypothesis-driven or question-driven. In other words, let the data reveal the information that they contain while we explore them without being guided by specific and focused questions. Indeed, this requires computational methods with exploratory capabilities.

One class of exploratory methods is unsupervised clustering. However, existing clustering and consensus clustering methods, such as the k-means, hierarchical clustering (HC) and relabelling and voting, are generally applicable to a single dataset, and even tend to generate different results when applied to the same dataset. This leaves the exploration of multiple heterogeneous datasets together quite unresolved.

We have taken steps to realise the data-driven paradigm of research more efficiently by introducing a consensus clustering method, the binarisation of consensus partition matrices (Bi-CoPaM) [2]. This method not only considers the consensus of the applications of different individual clustering methods, which is what existing consensus clustering methods do, it also has the ability to analyse multiple heterogeneous datasets collectively with the objective of identifying the subsets of objects which are collectively correlated in all of the given datasets. The datasets that the Bi-CoPaM receives should measure some features for the same set of objects. However, remarkably, the types and numbers of features in these individual datasets do not have to be the same neither does the technology used have to be the same. We also present two case studies in which we demonstrate the ability of such exploratory approach in finding novel conclusions of interest to the specific research communities relevant to the analysed datasets.

The first case study presents the analysis of forty different yeast gene expression datasets from a broad range of conditions such as the yeast cell-cycle, metabolic cycle and stress conditions. However, what is common to all of these forty datasets is that they are from the same species, the budding yeast *Saccharomyces cerevisiae*. Therefore, this exploratory experiment aims at discovering findings that are generally applicable to this species regardless of the particular conditions which it experiences. Indeed, the selection of the collection of datasets had a major role in defining that aim. The experiment identified two clusters of genes, one of them is novel, and few hypotheses regarding the regulation and control of core biological processes in yeast, namely, ribosome biogenesis and stress response [4].

The second case study presents the analysis of a large fMRI dataset which semantically consists of 1856 datasets, each one of them is a big dataset itself. Again, we demonstrate the exploratory power of our Bi-CoPaM method in extracting field-specific conclusions. Three functionally connected neural networks are identified related to visual, reward, and auditory systems during affective processing; this is one of the few studies that have observed these networks using a data-driven approach.

2 Methods

2.1 Bi-CoPaM

The Bi-CoPaM method is a consensus clustering method which can be applied to a set of datasets that quantify various sets of features for the same set of objects. The number and type of features (e.g. time points, conditions or stages) in these datasets does not have to be similar, as long as these datasets quantify some features for the same set of objects. The objective of the Bi-CoPaM is to identify the subsets (clusters) of objects which are consistently correlated with each other in all given datasets. Many objects may be left unassigned to any of the clusters if they do not meet the relevant external specification.

The Bi-CoPaM method consists of four main steps detailed in [2] and is summarised here:

1. Individual partitions generation
   Various clustering methods are independently applied to each one of the datasets. For M different methods and L different datasets, this step generates $C = M \times L$ partitions (clustering results). All of these clustering methods should be applied to generate the same number of clusters (K).
2. Relabelling
   Because clustering methods are unsupervised and their resulting clusters are unlabelled, the ith cluster of one of the C partitions may not necessarily correspond to the ith cluster of the other partitions. This issue is resolved by the relabelling step as it reorders the clusters in the partitions to be aligned. The min-max [2] and the min-min [4] techniques were proposed to realise the relabelling step, and the latter is what has been considered in the case studies in this paper.
3. Fuzzy consensus partition matrix (CoPaM) generation
   The relabelled partitions are combined by membership averaging to produce a fuzzy consensus partition matrix (CoPaM). The CoPaM associates each object with each cluster with a fuzzy membership value between zero and unity, where a zero membership means the object does not belong to the cluster, a unity membership means it fully belongs to it, and a value between zero and one indicates a partial belongingness.
4. Binarisation
   The fuzzy CoPaM is binarised by one of six techniques defined in [2] to assign objects to clusters with binary (crisp) values, i.e. an object either belongs or does not belong to any given cluster. Out of the six techniques, the technique which was adopted in all of the case studies here is the *difference threshold binarisation (DTB)* technique, which leaves the objects that do not significantly belong to a single cluster unassigned from all clusters. A tuning parameter $\delta \in [0, 1]$ controls the strictness of the DTB technique, which results in tighter but small clusters when increased, and results in larger but looser clusters when decreased.

2.2 M-N Scatter Plots Technique

As can be seen from the description of the Bi-CoPaM method above, two parameters need to be set, namely the number of clusters (K) and the tightness tuning parameter δ. Varying these parameters result in different results with clusters that vary widely in their size and tightness. The M-N scatter plots technique addresses this aspect by selecting the few clusters that best fulfill some quality criteria amongst all of those clusters generated by multiple applications of Bi-CoPaM to the same datasets with various values of K and δ.

The M-N scatter plot is a 2D plot on which each cluster is represented by a single point [5]. Its horizontal axis represents the dispersion (looseness) of the clusters measured by a modified version of the mean-squared error (MSE), while the vertical axis represents the size of the clusters measured by a scaled value of the logarithm of the number of objects included in the clusters.

The interest here is to select larger clusters with minimum dispersion, i.e. with greater tightness. Therefore, better clusters are those which maximise their value on the vertical axis while minimising it on the horizontal axis. In other words, better clusters are those which are closest to the top-left corner of the M-N scatter plot.

Consequently, the cluster whose point is closest to the top-left corner of the plot in Euclidean distance is selected first as the best cluster. Note that the clusters scattered on the plot are from various applications of the Bi-CoPaM over the same datasets, and thus many of them largely overlap. Therefore, all of the clusters which overlap with that first selected cluster are removed from the scatter plot. Then, out of the clusters remaining in the plot, the closest one to the top-left corner is selected as the second best cluster and all of those which overlap with it are removed from the plot. This iterative process is continued until no more clusters are left in the plots.

The result of this technique is a set of clusters listed in a decreasing order of their quality as measured by their distances from the top-left corner of the plot. In many applications, the first few clusters are considered meaningful while the rest of them are discarded. This can be decided based on the distribution of their distances and the specific requirements of the application.

3 Case Study I—Anti-correlated Gene Expression Profiles with Clusters of Yeast Genes

3.1 Summary

Eukaryotes are species with cells that have real nuclei and organelles, including all animals, plants, fungi and protists. Budding yeast (*S. cerevisiae*) is the most studied unicellular eukaryotic organism which has emerged as a model organism because of its easy manipulation and short cell-cycle, making it feasible to run rapid functional experiments. Many basic biological processes are shared by all eukaryotes; thus, understanding the biology of yeast can help in understanding the biology of many other species, including humans.

To identify the core subsets of genes which are globally co-expressed (correlated in gene expression) in yeast, we applied the Bi-CoPaM to forty different gene expression datasets that span various conditions and were generated in different years and laboratories. This study, published in *BMC Bioinformatics* [4], identified two core clusters of genes which are consistently co-expressed over all forty datasets. More interestingly, the expression of those two globally co-expressed clusters were negatively correlated in all of the datasets. This observation strongly suggests that a common machinery may be regulating both clusters of genes globally but in opposite directions, i.e. while this regulatory machinery activates one of the two clusters of genes, it represses the genes in the other cluster. Furthermore, it was observed that the first cluster always goes up in expression under growth conditions and goes down under stress conditions, while the second cluster behaves oppositely. As expected, the first cluster was found to be significantly enriched with growth-related genes, particularly rRNA processing and ribosome biogenesis. However, the second cluster included many genes with unknown functions, which we predicted to have roles in stress response. Finally, we deduced hypotheses regarding the regulatory machinery which may be regulating both clusters in opposite directions.

The rest of this section describes the datasets and the experiments in more detail.

3.2 Forty Genome-Wide Yeast Datasets from Various Conditions

Table 1 lists the forty budding yeast datasets which were submitted to the Bi-CoPaM method alongside some description of them. It is clear that the datasets were generated in different years, by different authors, under different conditions, and with different numbers of features. The specific biological contexts to which these datasets belong include the cell-cycle, the metabolic cycle and various stresses. However, each one of these datasets has the expression profiles of the same 5,667 yeast genes, which represent the entire yeast genome as represented by the Affymetrix yeast genome 2.0 array.

3.3 Two Globally Anti-correlated Clusters of Genes Identified in Yeast

The Bi-CoPaM method identified two clusters of genes which were, respectively, named as C1 and C2, and which , respectively, include 257 and 47 genes out of the initial 5,667 genes. The average expression profiles of these genes in each one of the forty datasets are shown in Fig. 1.

Table 1 Budding yeast gene expression datasets

ID	GEO accession*	Year	N^	Description	References
D01	GSE8799	2008	15	Two mitotic cell-cycles (w/t)	Orlando et al. [41]
D02	GSE8799	2008	15	Two mitotic cell-cycles (mutated cyclins)	Orlando et al. [41]
D03	E-MTAB-643$	2011	15	Response to an impulse of glucose	Dikicioglu et al. [18]
D04	E-MTAB-643$	2011	15	Response to an impulse of ammonium	Dikicioglu et al. [18]
D05	GSE54951	2014	6	Response of *dal80Δ* mutant yeast to oxidative stress induced by linoleic acid hydroperoxide	-#
D06	GSE25002	2014	9	Osmotic stress response and treatment of transformants expressing the *C. albicans* Nik1 gene	-#
D07	GSE36298	2013	6	Mutations of OPI1, INO2 and INO4 under carbon-limited growth conditions	Chumnanpuen et al. [15]
D08	GSE50728	2013	8	120-h time-course during fermentation	-#
D09	GSE36599	2013	5	Stress adaptation and recovery	Xue-Franzén et al. [58]
D10	GSE47712	2013	6	Combinations of the yeast mediator complex's tail subunits mutations	Larsson et al. [32]
D11	GSE21870	2013	4	Combinations of mutations in DNUP60 and DADA2	-#
D12	GSE38848	2013	6	Various strains under aerobic or anaerobic growth	Liu et al. [35]
D13	GSE36954	2012	6	Response to mycotoxic type B trichothecenes	Suzuki and Iwahashi [51]
D14	GSE33276	2012	6	Response to heat stress for three different strains	Nakamura et al. [40]
D15	GSE40399	2012	7	Response to various perturbations (heat, myriocin treatment and lipid supplement)	Montefusco et al. [37]
D16	GSE31176	2012	6	W/t, *rlm1Δ* and *swi3Δ* cells with or without Congo Red exposure	Sanz et al. [46]
D17	GSE26923	2012	5	Varying levels of GCN5 F221A mutant expression	Lanza et al. [31]
D18	GSE30054	2012	31	CEN.PK122 oscillating for 2 h	-#
D19	GSE30051	2012	32	CEN.PL113-7D oscillating for 2 h	Chin et al. [14]
D20	GSE30052	2012	49	CEN.PL113-7D oscillating for 4 h	Chin et al. [14]
D21	GSE32974	2012	15	About 5 h of cell-cycle (w/t)	Kovacs et al. [30]

(continued)

Table 1 (continued)

ID	GEO accession*	Year	N^	Description	References
D22	GSE32974	2012	15	About 4 h of cell-cycle (mutant lacking Cdk1 activity)	Kovacs et al. [30]
D23	GSE24888	2011	5	Untreated yeast versus yeasts treated with *E. arvense* herbs from the USE, China, Europe or India	Cook et al. [16]
D24	GSE19302	2011	6	Response to degron induction for w/t and nab2-td mutant	González-Aguilera et al. [24]
D25	GSE33427	2011	5	Untreated w/t, and wt/t, *yap1Δ*, *yap8Δ* and double mutant treated with AsV	Ferreira et al. [21]
D26	GSE17716	2011	7	Effect of overexpression and deletion of MSS11 and FLO8	Bester et al. [10]
D27	GSE31366	2011	4	Presence and absence of mutli-inhibitors for parental and tolerant strains	-#
D28	GSE26171	2011	4	Response to patulin and/or ascorbic acid	Suzuki and Iwahashi [50]
D29	GSE22270	2011	4	PY1 and Met30 strains in room temperature or 35 °C	-#
D30	GSE29273	2011	4	Time-series during yeast second fermentation	-#
D31	GSE29353	2011	5	Different haploid strains growing in low glucose medium	Parreiras et al. [42]
D32	GSE21571	2011	8	Different combinations of mutations in HTZ1, SWR1, SWC2 and SWC5	Morillo–Huesca et al. [39]
D33	GSE17364	2010	4	Untreated w/t and Slt2-deficient yeasts, or treated with sodium arsenate for 2 h	Matia-González and Rodríguez-Gabriel [36]
D34	GSE15352	2010	8	24-h time-course of yeast grown under a low temperature (10 °C)	Strassburg et al. [48]
D35	GSE15352	2010	8	24-h time-course of yeast grown under a normal temperature (28 °C)	Strassburg et al. [48]
D36	GSE15352	2010	8	24-h time-course of yeast grown under a high temperature (37 °C)	Strassburg et al. [48]
D37	GSE16799	2009	21	UC-V irradiation of w/t, *mig3Δ*, *SNF1Δ*, *RAD23Δ*, *RAD4Δ* and *snf1Δrad23Δ*	Wade et al. [56]
D38	GSE16346	2009	4	BY474 cells grown to mid-log under presence versus absence of L-carnitine and/or H_2O_2	-#

(continued)

Table 1 (continued)

ID	GEO accession*	Year	N^	Description	References
D39	GSE14227	2009	10	Two hours of wild-type yeast growth	[23]
D40	GSE14227	2009	9	Two hours of *sch9Δ* mutant yeast growth	[23]

*GEO accession: the unique accession number of the dataset in the GEO repository. All of these datasets are publically available at this repository
$This particular accession is not a GEO accession number; it is rather an ArrayExpress accession number which can be used to access this dataset at the EMBL-EBI ArrayExpress repository
^NThe number of features (biological conditions or time points) which the dataset covers
#No references are available for some datasets. However, they can still be accessed on the GEO repository through their GEO accession

Interestingly, C1 and C2 show negative correlation consistently in each of the 40 datasets. Correlation values between the average expression profiles of C1 and C2 in each one of the datasets are shown in Fig. 2, where it can be seen that in 38 datasets the correlation value does not exceed the −0.70, and that it is still a negative (but less significant) value in the two other datasets. This suggests that both clusters may be regulated by a common regulatory machinery.

3.4 *C1 is a Ribosome Biogenesis (RiB) Cluster While C2 is Predicted to be a Stress Response Cluster*

Gene ontology (GO) term analysis [47], at the time of our original study [4], showed that C1 is enriched with genes known to participate in ribosome biogenesis (p-value = 10^{-129}). Ribosome biogenesis is required to build the ribosomes which synthesise proteins, and are therefore largely needed under growth conditions. By scrutinising the conditions under which the expression of C1 goes up across all 40 datasets, it can be clearly seen that expression of genes in the C1 cluster does go up under growth conditions, as predicted from the gene ontology analysis.

On the other hand, C2, which goes down under growth conditions and up under stress conditions, included some stress response genes (only 6 out 47 with a p-value of 10^{-3}), and many genes with unknown or unrelated functions. We labelled this cluster as "anti-phase with ribosome biogenesis" (APha-RiB). Given these observations, we predicted that many more genes in C2 may have roles in stress response.

It is worth pointing out that a comparison with other sets of genes identified by other studies as up-regulated under stress or down-regulated under growth demonstrated that C2 is largely a novel cluster of genes with many genes previously not identified as such [4]. This is despite the fact that C2 is much smaller than those

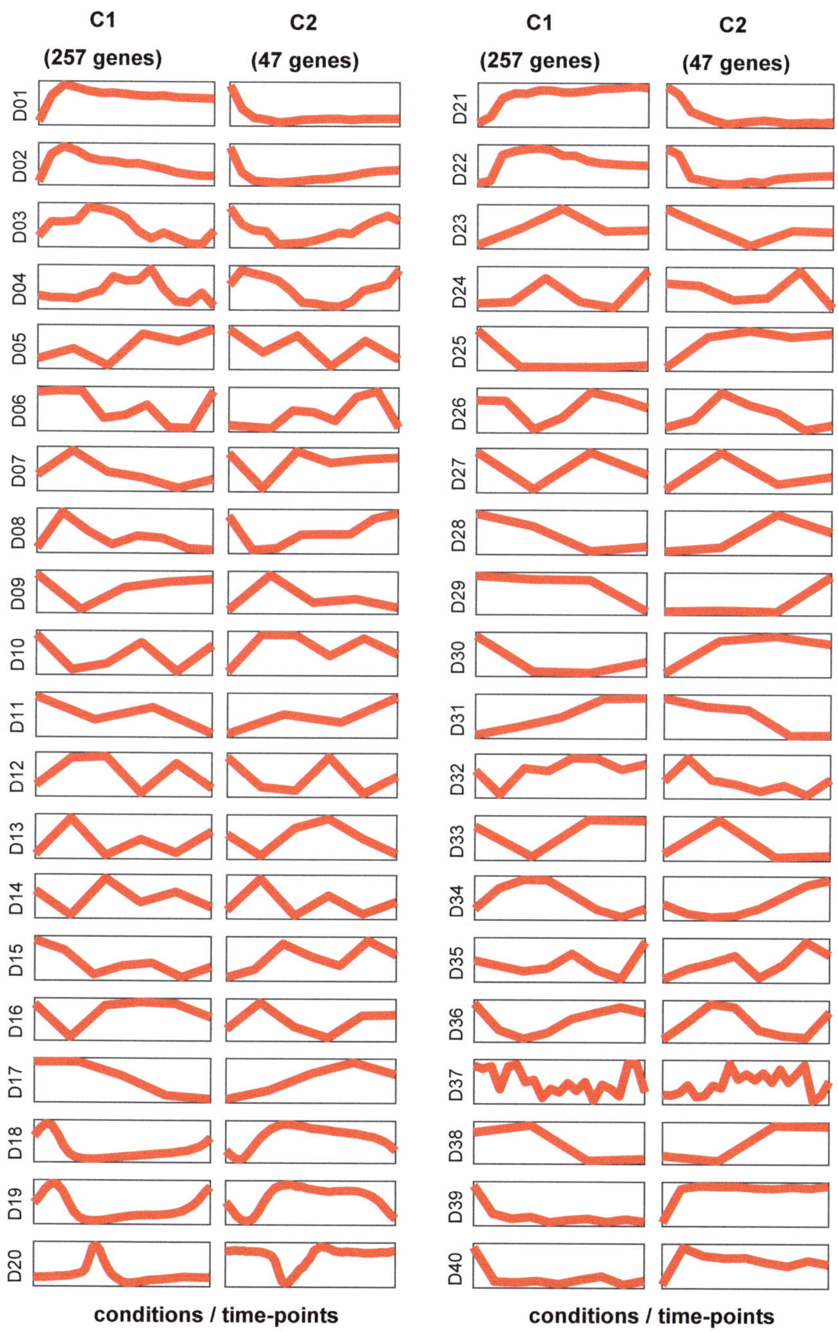

Fig. 1 The average expression profiles of the genes in clusters C1 and C2 over each one of the forty datasets D01 to D40

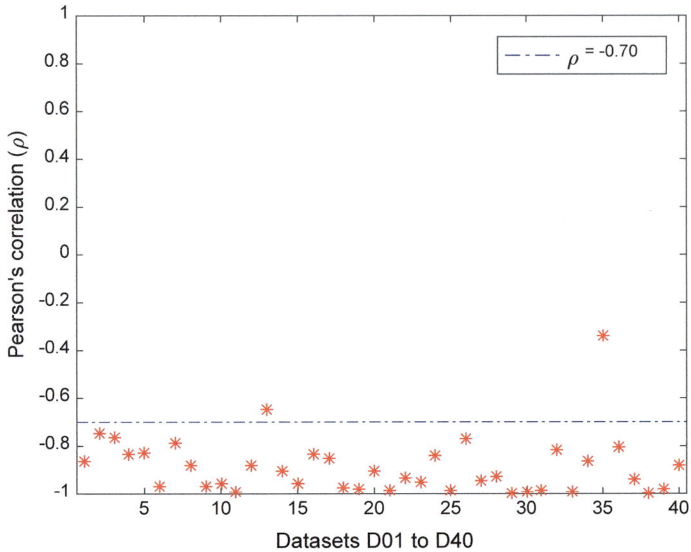

Fig. 2 Pearson's correlation values between the average profiles of C1 and C2 in each one of the forty datasets. The *horizontal dotted line* marks the correlation value of −0.70

other sets which include between 280 and 550 genes. Indeed, identifying a focused cluster of genes which is consistently anti-correlated with growth in forty datasets from broad conditions is more significant than identifying a large cluster of genes which are anti-correlated with growth in a single study. Many genes in the latter might well show such behaviour out of coincidence, while those in the former are highly expected to be genuinely controlled by growth and stress conditions. This demonstrates the power of our approach in identifying clusters of genes from collections of datasets.

3.5 Predicted Regulatory Network Controlling C1 (RiB) and C2 (APha-RiB)

We used the MEME [8] and the TOMTOM [25] tools to identify the regulatory elements (DNA motifs) and the regulators that may control C1 (RiB) and C2 (APha-RiB). Figure 3 summarises our findings, together with previous observation from the literature, in the form of a regulatory network.

Many of the regulatory interactions that we predicted agreed with what has been found in the literature. However, some other interactions were not recognised before our study and we hypothesised that they may exist. Notably, the transcription factor (TF) Stb3 was shown to repress RiB genes under stress, and we

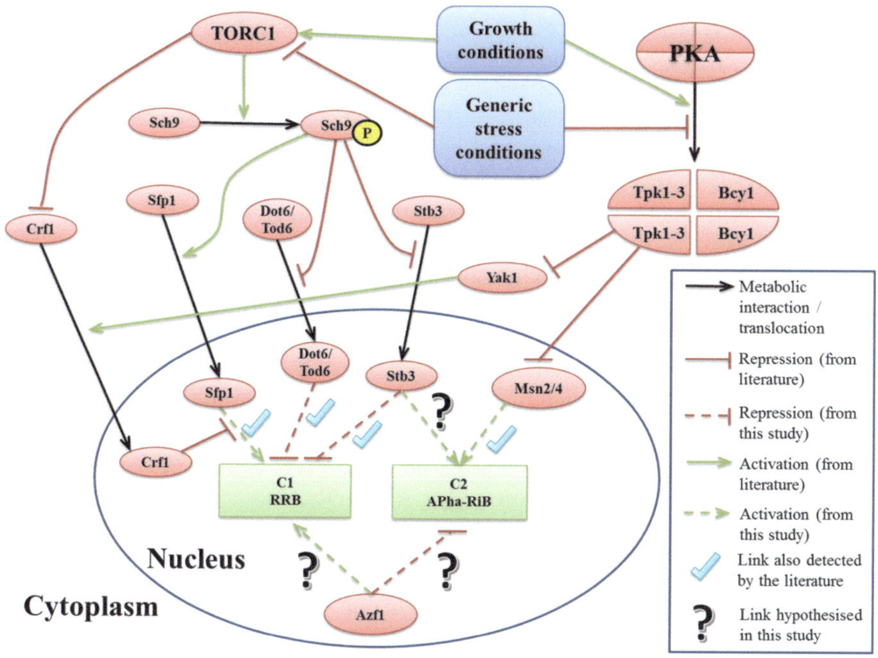

Fig. 3 Predicted regulatory network that controls both C1 and C2 while showing the regulatory interactions predicted in this study and the ones which were derived from the literature

predicted that it may have a role in activating APha-RiB genes at the same time. Also, we predicted that the regulator Azf1 may have a role in activating RiB genes while repressing APha-RiB genes under growth conditions.

4 Case Study II—Brain Functional Network Involved in Affective Processing

4.1 *Background*

Blood oxygen level dependent (BOLD) signals measured by fMRI datasets indirectly reflect the neural activities during cognitive tasks [22, 38] or during resting-state [33]. Obtaining these brain data is the first and fundamental step to capture the brain status. To further study the brain functions, information conveyed by brain data needs to be extracted and quantitatively analysed. Up to now, various fMRI data analysis methods have been developed such as statistical parametric mapping (SPM) [22], seed-based connectivity analysis, independent component analysis (ICA), multivariate pattern analysis (MVPA) [17] as well as clustering analysis [55]. It is difficult to state which method outperforms others, rather the choice of method depends on the design of fMRI experiment paradigm.

Traditional SPM method has been used for decades. The modelling of BOLD signal suits the fMRI experiments with stimuli having explicit timing and duration but when it comes to the resting-state, i.e. no stimuli are given during scanning, or very complex experiment paradigm, then modelling of these BOLD signals becomes very difficult or impossible. Recently, a study reported that many modern software packages incorporating SPM method for analysing neuroimaging data suffer from "bugs" that could invalidate many statistical results [19]. Seed-based connectivity analysis searches the voxels showing similar BOLD fluctuations to the predefined seed area/voxel, which limits its application to regional analysis rather than the whole brain scale connectivity analysis.

ICA can decompose the observed BOLD signals into a set of independent sources/components, which could be mixed to reconstitute the observed signals. However, the underlying assumption that the components are mutually independent is debatable. Also the components yielded by the analysis are not easy to interpret and still less validated. MVPA is based on supervised classification that detects differences between conditions with higher sensitivity than conventional univariate analysis (e.g. SPM) by focusing on the analysis and comparison of distributed patterns of activity (multivariate), but it also only suits the fMRI experiments with known and well separated stimuli due to the supervised classification nature. Clustering approach, which belongs to unsupervised classification, attempts to divide the data objects into clusters where objects in the same cluster are more similar to each other than objects from different clusters. The unsupervised feature makes clustering a very flexible and powerful tool of exploring and analysing the neuroimaging data. It has gained a lot of popularity in connectivity analysis and brain parcellation [34, 52].

Clustering itself is an ill-posed problem where the solution violates at least one of the common assumptions regarding cluster consistency, richness and scale invariance [26]. Almost every clustering algorithm explicitly or implicitly assumes a certain data model based on which the data is clustered [49]. Without a priori knowledge, each clustering algorithm could produce plausible results based on its underlying assumption of the data structures that might not be satisfied by the input data. To match the assumed data model and the true input data structure, it is crucial to obtain the correct a priori information to obtain good clustering results; however, such a priori information is often hard to obtain even from experts. A recent main criticism from the neuroscience field relies on the wide variety of analysis strategies, combined with small sample sizes, used to investigate regional brain activity measured with fMRI that frequently lead to inconsistent findings [13].

To answer these common problems in clustering analysis, we explored the consensus clustering analysis (Bi-CoPaM) of fMRI data and its benefits compared to single clustering methods such as K-means, hierarchical clustering and self-organising map (SOM). The validation results show that there are obvious differences among the clustering solutions by different single clustering algorithms as well as the different combination of single algorithms. Bi-CoPaM paradigm can not only find the objects that consistently belong to same cluster across many clustering experiments but also is able to extract the clusters according to their tightness (quality) through tuning the consensus level. These selected non-trivial clusters have very strong inner similarity as well as relatively large number of members, making it more reliable to draw neuroscientific conclusions.

4.2 fMRI Experiment

The fMRI data comes from an experiment studying human brain responses to music with different preference and emotions. All the music stimuli are divided into four categories that are liked happy (LH) [this represents the case of when a participant considers the music is "happy" and "likes" it], liked sad (LS) [this represents the case of when a participant considers the music is "sad" and "likes" it], disliked happy (DH) [this represents the case of when a participant considers the music is "happy" and "dislikes" it], and disliked sad (DS) [this represents the case of when a participant considers the music is "sad" and "dislikes" it]. The participants were from two groups that are musicians, who have received systematically music performance training, and non-musicians. A total of 13 musicians and 16 non-musicians participated in the fMRI scanning sessions. The whole fMRI experiment for one participant consists of 450 whole brain scans (TR = 3 s) while listening to one of the 32 music categories with each one repeated twice. Each scan covering the whole brain contains 228,453 voxels after pre-processing. In total, there are 1856 fMRI data segments corresponding to 1856 (= 29 × 32 × 2) listening trials. The study was approved by the ethical committee of the Helsinki University Central Hospital and complied with the Helsinki Declaration. The dataset is a subset of a larger data collection, parts of which have been published in [11, 44].

4.3 Clustering Experiment

The K-means, Hierarchical and SOM were applied on each of 1856 excerpts data (normalised to 0 mean and unit variance) with K equals to 10, 25, 50 and 100. These clustering results generated by different algorithms with four different cluster numbers were combined using the Bi-CoPaM paradigm for each possible combination of three methods, followed by cluster ranking and selection by the M-N scatter plot technique, yielding the consensus clustering results. In total, seven sets of final clustering results were obtained, i.e. clustering results from K-means, hierarchical clustering, SOM, K-means and hierarchical clustering, K-means and SOM, hierarchical clustering and SOM, and a combination of all the three methods. The brain regions within clusters were extracted by using automated anatomical labelling (AAL) atlas [54] and names of these regions were used to identify the neuroscience background of clusters.

4.4 Comparing Clustering Results Among Different Combination Scenarios

We chose the clusters covering three important areas in human brain, namely the auditory system, reward system and visual system, to evaluate the accuracy and completeness of clustering results. To measure quantitatively the similarity between two clusters, Jaccard index is used. Table 2 shows the Jaccard index between different method combinations for the cluster covering visual area. Table 3 shows the Jaccard index between different method combinations for the cluster covering reward system and Table 4 shows the Jaccard index between different method combinations for the cluster covering auditory system. KM represents K-means and HC represents hierarchical clustering. Figures 4, 5 and 6 are the visualisations of different clustering results.

Note that for those methods or method combinations having all zero Jaccard index with other clusters means that this method or method combination was not able to detect the corresponding cluster. For example, in Table 4, SOM method did not group the voxels within the auditory system into a cluster; neither did the combination of KM and HC. The combination of all the three methods never failed to detect any cluster.

4.5 Robustness Test Against Subject Functional Data Variability

Considering that the brain activity to a certain stimulus varied among different individuals, it raises the problem whether the algorithm could perform well regarding the functional data variation among participants. Since it has been demonstrated that Bi-CoPaM with three methods provides the most complete set of clusters, we carried out the test to investigate the robustness of Bi-CoPaM with three clustering methods against data variability on afore-mentioned three clusters. We generated two groups of subsets for test. One was created by randomly selecting 75% of the musicians (10 out of 13) and non-musicians (12 out of 16) as well as 75% of the excerpts for each participant, which yields a subset consisting approximately 56% of all the data from the fMRI experiment. The above random selection was repeated 10 times and these 10 subsets formed group A. Similarly, we chose a different ratio of 90% of the musicians (12 out of 13) and non-musicians (14 out of 16) as well as 90% of the excerpts for each participant, when the data were randomly selected. Repeating the selection with new ratio for 10 times formed group B consisting approximately 80% of all the data. Then we applied Bi-CoPaM with three clustering algorithms on these subsets and recorded the clustering results.

From Tables 5 and 6, we could see, in most of the trials, the three important clusters (Visual, Reward and Auditory) could be identified despite different subsets being used. Meanwhile, with the increasing proportion of data used, the results

Table 2 Jaccard index for the cluster covering visual area

	KM	HC	SOM	KM + HC	KM + SOM	HC + SOM	KM + HC + SOM
KM		0.48	0.70	0.70	0.78	0.73	0.40
HC	0.48		0.38	0.57	0.53	0.56	0.26
SOM	0.70	0.38		0.58	0.66	0.62	0.50
KM + HC	0.70	0.57	0.58		0.76	0.84	0.32
KM + SOM	0.78	0.53	0.66	0.76		0.82	0.37
HC + SOM	0.73	0.56	0.62	0.84	0.82		0.35
KM + HC + SOM	0.40	0.26	0.50	0.32	0.37	0.35	

Table 3 Jaccard index for the cluster covering reward system

	KM	HC	SOM	KM + HC	KM + SOM	HC + SOM	KM + HC + SOM
KM		0.00	0.56	0.41	0.53	0.52	0.53
HC	0.00		0.00	0.00	0.00	0.00	0.00
SOM	0.56	0.00		0.43	0.57	0.55	0.57
KM + HC	0.41	0.00	0.43		0.33	0.33	0.37
KM + SOM	0.53	0.00	0.57	0.33		0.67	0.61
HC + SOM	0.52	0.00	0.55	0.33	0.67		0.60
KM + HC + SOM	0.53	0.00	0.57	0.37	0.61	0.60	

Table 4 Jaccard index for the cluster covering auditory system

	KM	HC	SOM	KM + HC	KM + SOM	HC + SOM	KM + HC + SOM
KM		0.49	0.00	0.00	0.41	0.71	0.62
HC	0.49		0.00	0.00	0.35	0.51	0.56
SOM	0.00	0.00			0.00	0.00	0.00
KM + HC	0.00	0.00			0.00	0.00	0.00
KM + SOM	0.41	0.35	0.00	0.00		0.48	0.40
HC + SOM	0.71	0.51	0.00	0.00	0.48		0.65
KM + HC + SOM	0.62	0.56	0.00	0.00	0.40	0.65	

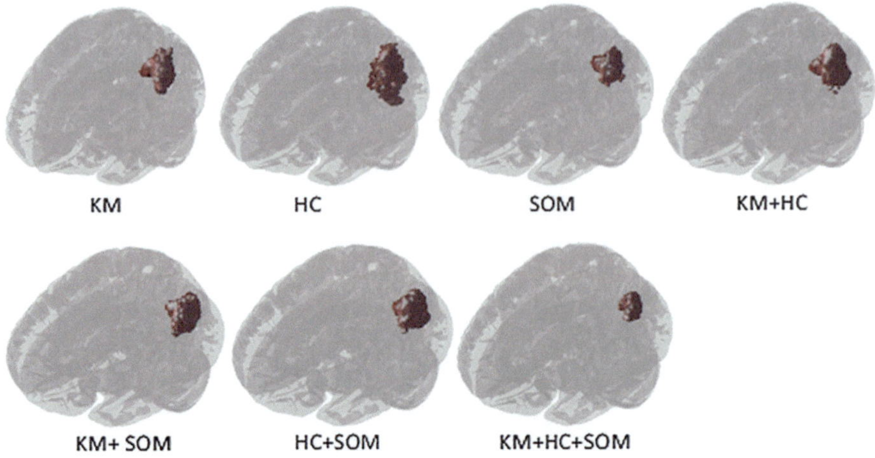

Fig. 4 3D illustrations of clusters covering visual area detected by seven different method combinations

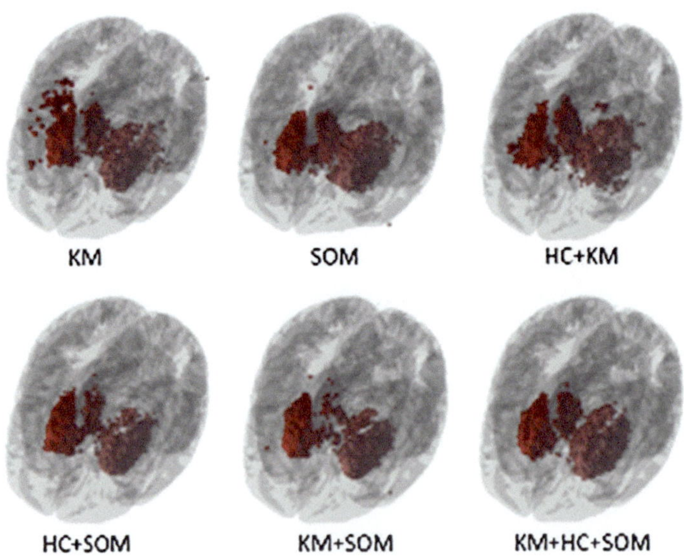

Fig. 5 3D illustrations of clusters covering reward system detected by six different method combinations

become more stable. For example, the Reward cluster was missed three times in the first experiment reported in Table 5, but not missed at all in the experiment reported in Table 6.

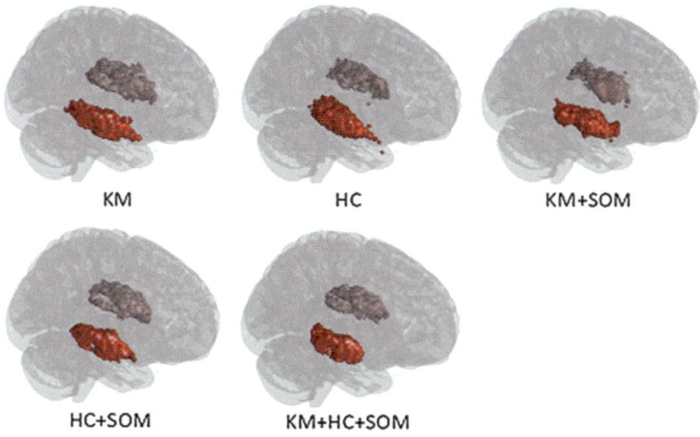

Fig. 6 3D illustrations of clusters covering auditory system detected by five different method combinations

4.6 *Topology of Final Clusters*

From Sects. 4.4 and 4.5, it can be concluded that the consensus scenario with all the three clustering algorithms gives relatively the best solution among all the possibilities of method combination. Figure 7 illustrates the clusters from combining all the three clustering algorithms in details such as their topologies within anatomical brain structures and the area information within each cluster.

Cluster A comprises bilateral visual areas, namely the calcarine fissure and the cuneus. Cluster B comprises bilateral neural structures of the reward system, namely the ventral striatum—extending to the globus pallidus—the thalamus, the amygdala, the orbitofrontal cortex and the left insula. Cluster C comprises the auditory areas, namely the bilateral superior temporal gyrus, Heschl's gyrus, the left middle temporal gyrus, as well as one region of the somatosensory cortex, namely the right Rolandic operculum. These areas correspond well with the model-based literature [11].

4.7 *Summary of Case Study II*

In this study, Bi-CoPaM was used to combine the clustering results on a real fMRI data with multiple clustering algorithms to improve the consistency and stability performance of clustering in neuroimaging data analysis. The results show that the proposed method has great capability and offer much potential to help address the consistency issue in clustering analysis of neuroimaging data.

Table 5 Results of test group A (56% of the full data). Final size is the obtained cluster size using the whole participants' data. Trial size is the cluster size obtained in different trials. Intersection is the size of the part that the clusters in each trial intersect with the clusters obtained using the whole participants' data

Clusters		Trial										Mean
		1	2	3	4	5	6	7	8	9	10	
Visual	Final size	410	410	410	410	410	410	410	410	410	410	
	Trial size	551	1031	961	900	924	1838	713	877	2027	651	
	Intersection	357	410	410	410	405	410	396	409	410	393	
	Dice Coeff.	0.74	0.57	0.60	0.63	0.61	0.36	0.71	0.64	0.34	0.74	0.594
Reward	Final size	4293	4293	4293	4293	4293	4293	4293	4293	4293	4293	
	Trial size	2857	2591	4646	0	3263	1028	0	1620	0	4168	
	Intersection	1479	2046	3218	0	2278	915	0	1403	0	3303	
	Dice Coeff.	0.41	0.59	0.72	0	0.60	0.34	0	0.47	0	0.78	0.391
Auditory	Final size	2123	2123	2123	2123	2123	2123	2123	2123	2123	2123	
	Trial size	2381	3477	1452	2980	1576	2191	1801	2161	856	2006	
	Intersection	1738	2021	16	1995	439	1581	1459	1343	797	1673	
	Dice Coeff.	0.77	0.72	0	0.78	0.24	0.73	0.74	0.63	0.54	0.81	0.596

Table 6 Results of test group B (80% of the full data)

Clusters		Trial										Mean
		1	2	3	4	5	6	7	8	9	10	
Visual	Final size	410	410	410	410	410	410	410	410	410	410	
	Trial size	591	991	385	821	803	1119	784	1216	821	803	
	Intersection	402	410	326	406	406	410	407	410	406	406	
	Dice Coeff.	0.80	0.59	0.82	0.66	0.67	0.54	0.68	0.50	0.66	0.67	0.66
Reward	Final size	4293	4293	4293	4293	4293	4293	4293	4293	4293	4293	
	Trial size	3028	2867	3118	4381	1813	4390	4426	2362	4381	1813	
	Intersection	1273	1027	2467	3308	1519	3157	3461	1891	3308	1519	
	Dice Coeff.	0.35	0.29	0.67	0.76	0.50	0.73	0.79	0.57	0.76	0.50	0.592
Auditory	Final size	2123	2123	2123	2123	2123	2123	2123	2123	2123	2123	
	Trial size	476	1948	2910	1348	2496	3158	3036	771	1348	2496	
	Intersection	458	1548	1997	338	1855	2051	1704	578	338	1855	
	Dice Coeff.	0.35	0.76	0.79	0.19	0.8	0.78	0.66	0.40	0.19	0.8	0.572

From Tables 2, 3 and 4, which show the mutual differences between any two method combination scenarios, and the three-dimensional demonstration of all the clusters covering three important neural systems in human brain, we can infer the obvious differences among different clustering schemes. For the cluster covering the visual system, although all of the method combinations were able to detect this area, details of the regions such as the size and shape vary among each other. For the cluster covering the reward system and another cluster covering the auditory system, not only the results are different from each combination of clustering algorithms but also some single methods failed to detect the important brain structures based on their BOLD response time profiles. However, when it comes to the results obtained from any combined results from different single methods, these two important clusters could be identified, and only one scenario (the combination of KM and HC) failed to form the auditory cluster. This demonstrates the power of integrating the results from multiple clustering methods. As for the individual functional data variability test, Tables 5 and 6 demonstrate that Bi-CoPaM can detect the three important clusters in most of the trials consisting only part of all the data. On one hand, this demonstrates that the Bi-CoPaM is robust to variability of participants' data and thus generates reproducible results; on the other hand, it also shows the benefits of using a large number of subjects for more reliable results in data-driven analysis of functional brain imaging data.

From the neuroscientific aspect, the innovative use of Bi-CoPaM allows us to find clusters including functionally and anatomically related neural networks responding to emotional music. After the cluster generation and selection, emotion-related brain structures responsible for rewarding and pleasurable sensations such as the basal ganglia, thalamus, insula [45] and other areas involved with processing of auditory features such as the Heschl's gyrus, the Rolandic operculum and the superior temporal gyrus [7] were grouped into corresponding clusters separately. One of the most important findings of this study is that, without any predetermined model assigning a value to each stimulus, the Bi-CoPaM was able to obtain a single cluster including the anatomically connected sub-cortical and cortical structures of the reward circuit, responding to music with various emotions and preferences. This is one of the few studies obtaining such finding with a data-driven method. For example, a recent study using a data-driven network science method to study affective music processing [57], did not observe any reward circuit activity. As a matter of fact, neural structures of the reward circuit have also been found to be more or less connected only in other functional connectivity analysis studies such as one studying attention—deficit/hyperactivity disorder (ADHD) [53]. Our study confirms findings on the neural structures related to musical emotions obtained with model-based approaches [9, 11, 12, 27–29, 45].

Cluster A Visual

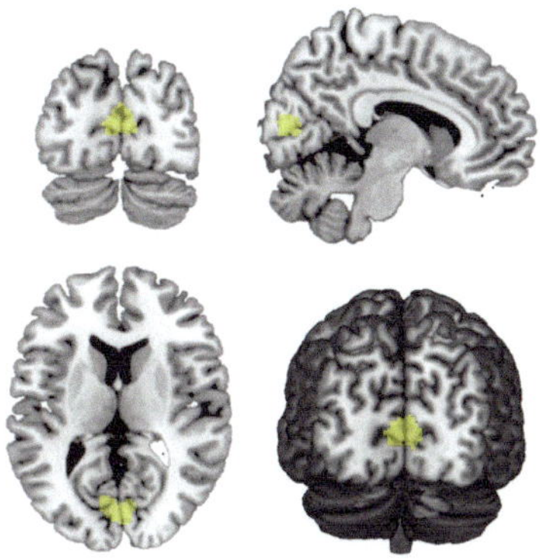

Area	Size (in voxel 2×2×2mm³)	MNI
Calcarine Fissure and surrounding cortex (L)	209	(-3 -77 11)
Calcarine Fissure and surrounding cortex (R)	95	(7 -78 11)
Cuneus (L)	70	(0 -80 19)
Cuneus (R)	36	(6 -81 18)

Cluster B Reward

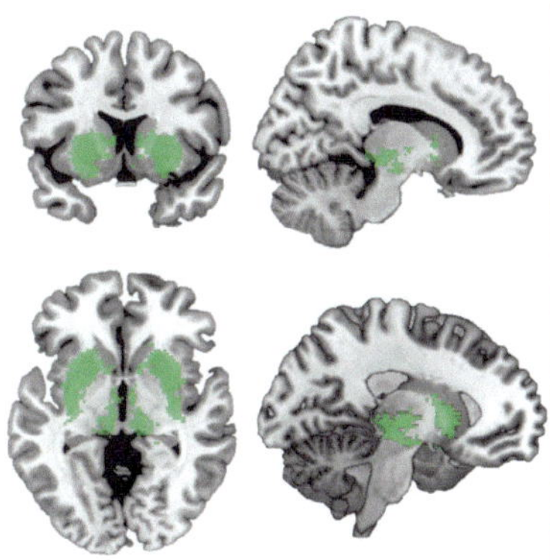

Area	Size (in voxel)	MNI
Putamen (L)	824	(-25 3 0)
Putamen (R)	870	(28 4 0)
Globus pallidus (L)	109	(-20 0 0)
Globus pallidus (R)	115	(22 0 1)
Thalamus (L)	168	(-11 -19 1)
Thalamus (R)	176	(14 -21 1)
Caudate nucleus (L)	116	(-12 13 3)
Caudate nucleus (R)	116	(16 16 3)
Amygdala (L)	50	(-25 -2 -15)
Amygdala (R)	37	(28 -2 -14)
Insula (L)	122	(-33 3 -2)
Olfactory cortex (L)	34	(-20 7 -15)
Olfactory cortex (R)	17	(23 10 -14)

Cluster C Auditory

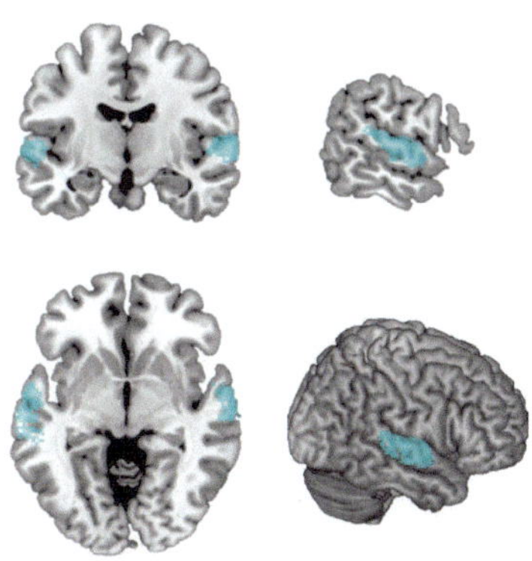

Area	Size (in voxel)	MNI
Middle temporal gyrus (L)	513	(-61 -28 0)
Superior temporal gyrus (L)	568	(-59 -16 4)
Superior temporal gyrus (R)	933	(61 -17 5)
Heschl's gyrus (L)	10	(-58 -12 8)
Heschl's gyrus (R)	42	(56 -9 6)
Rolandic operculum (R)	40	(60 -9 9)

Fig. 7 The 3D illustrations of clusters and the size of each sub-cluster with voxels falling within a known anatomical brain structure, identified with the AAL atlas

5 Discussion

Data-driven exploration of multiple datasets collectively has the ability to find novel results that are not straightforwardly accessible through the analysis of individual datasets and serves in the exploitation of the huge amounts of existing partially exploited datasets. The Bi-CoPaM paradigm [2] serves as a powerful example of a method which can be used to achieve such goals as demonstrated by the two case studies presented.

The two chosen case studies were selected from two distinct fields of research with datasets of different types, sizes and dynamic ranges. This implies that the scope of the method's applicability is much broader than the contexts of bioinformatics data or fMRI data; it is rather, expected to be applicable to many other fields, as long as their data were arranged in the proper format.

It may be argued that the methods which are only applicable to a single dataset can be applied to collections of datasets if the datasets themselves were combined, by concatenation for example, to form a single bigger dataset. Although this approach may work in some cases, it is not applicable when the datasets have key differences in their format, the technology used, or the dynamic ranges of their values. In some cases, certain normalisation and standardisation techniques can help in transforming the individual datasets into a unified format with dynamic ranges that make it valid to combine them, but in many other cases this cannot be done. For instance, gene expression datasets generated using one-colour microarrays, two-colour microarrays and next-generation sequencing cannot be trivially combined. Moreover, merging many datasets into a single dataset may result in an even larger single dataset that may not be feasibly analysed using the available computational power. In addition to that, the integrity of each dataset, as an independent dataset, sometimes is lost once it has been merged with other datasets. Taken together, the analysis of multiple datasets collectively can indeed be fruitfully achieved using the Bi-CoPaM paradigm even when they are heterogeneous and not trivially combinable.

Available datasets in the fields of bioinformatics and neuroscience are many more than what has been analysed by the Bi-CoPaM. Few other studies by the Bi-CoPaM have been already been published, such as the analysis of five *E. coli* bacterial datasets [6] and the analysis of two yeast cell-cycle datasets [3]. However, revisiting the collections of datasets from many other contexts is expected to yield new insights in biology and neuroscience as well as many other fields.

To conclude, we recommend that the large amounts of existing and emerging big datasets should be revisited with data-driven explorations to accelerate the pace of information inference. The Bi-CoPaM is indeed a method that has been shown to be able to achieve that goal in many applications and can be used widely to explore many more collections of datasets. Furthermore, efforts to design other sophisticated methods that can handle large amounts of heterogeneous datasets from other points of view, other than the Bi-CoPaM's, are required.

References

1. B. Abu-Jamous, R. Fa, A.K. Nandi, *Integrative cluster analysis in bioinformatics*, 1st edn. (Wiley, s.l., 2015)
2. B. Abu-Jamous, R. Fa, D.J. Roberts, A.K. Nandi, Paradigm of tunable clustering using binarization of consensus partition matrices (Bi-CoPaM) for gene discovery. PLoS ONE **8**(2), e56432 (2013)
3. B. Abu-Jamous, R. Fa, D.J. Roberts, A.K. Nandi, Yeast gene CMR1/YDL156W is consistently co-expressed with genes participating in DNA-metabolic processes in a variety of stringent clustering experiments. J. R. Soc. Interface **10**(81), 20120990 (2013)
4. B. Abu-Jamous, R. Fa, D.J. Roberts, A.K. Nandi, Comprehensive analysis of forty yeast microarray datasets reveals a novel subset of genes (APha-RiB) consistently negatively associated with ribosome biogenesis. BMC Bioinform. **15**, 322 (2014)
5. B. Abu-Jamous, R. Fa, D.J. Roberts, A.K. Nandi, UNCLES: method for the identification of genes differentially consistently co-expressed in a specific subset of datasets. BMC Bioinform. **16**, 184 (2015)
6. B. Abu-Jamous, R. Fa, D.J. Roberts, A.K. Nandi, Application of the Bi-CoPaM method to five Escherichia coli datasets generated under various biological conditions. J. Signal Process. Syst. **79**(2), 159–166 (2015)
7. V. Alluri et al., Large-scale brain networks emerge from dynamic processing of musical timbre, key and rhythm. NeuroImage **59**(4), 3677–3689 (2012)
8. T.L. Bailey, C. Elkan, *Fitting a mixture model by expectation maximization to discover motifs in biopolymers* (AAAI Press, Menlo Park, CA, s.n., 1994), pp. 28–36
9. L. Barrett, T. Wager, The structure of emotion evidence from neuroimaging studies. Curr. Dir. Psychol. Sci. **15**(2), 79–83 (2006)
10. M.C. Bester, D. Jacobson, F.F. Bauer, Many Saccharomyces cerevisiae cell wall protein encoding genes are coregulated by Mss11, but cellular adhesion phenotypes appear only Flo protein dependent. G3 (Bethesda) **2**(1), 131–141 (2012)
11. E. Brattico et al., A functional MRI study of happy and sad emotions in music with and without lyrics. Front. Psychol. **2**(December), 308 (2011)
12. E. Brattico, B. Bogert, T. Jacobsen, Toward a neural chronometry for the aesthetic experience of music. Front. Psychol. **4**, 206 (2013)
13. K.S. Button et al., Power failure: why small sample size undermines the reliability of neuroscience. Nat. Rev. Neurosci. **14**(5), 365–376 (2013)
14. S.L. Chin, I.M. Marcus, R.R. Klevecz, C.M. Li, Dynamics of oscillatory phenotypes in Saccharomyces cerevisiae reveal a network of genome-wide transcriptional oscillators. FEBS J. **279**(6), 1119–1130 (2012)
15. P. Chumnanpuen, I. Nookaew, J. Nielsen, Integrated analysis, transcriptome-lipidome, reveals the effects of INO-level (INO2 and INO4) on lipid metabolism in yeast. BMC Syst. Biol. **7** (Suppl 3), S7 (2013)
16. R. Cook et al., The Saccharomyces cerevisiae transcriptome as a mirror of phytochemical variation in complex extracts of Equisetum arvense from America, China, Europe and India. BMC Genomics **14**, 445 (2013)
17. F. De Martino et al., Combining multivariate voxel selection and support vector machines for mapping and classification of fMRI spatial patterns. NeuroImage **43**(1), 44–58 (2008)
18. D. Dikicioglu et al., How yeast re-programmes its transcriptional profile in response to different nutrient impulses. BMC Syst. Biol. **5**, 148–163 (2011)
19. A. Eklund, T. Nichols, H. Knutsson, Cluster failure: Why fMRI inferences for spatial extent have inflated false-positive rates. Proc. Natl. Acad. Sci. **113**(28), 7900–7905 (2016)
20. G. Elliott, A. Timmermann, Forecasting in economics and finance. Annu. Rev. Econ. **8**, 81–110 (2016)
21. R.T. Ferreira et al., Arsenic stress elicits cytosolic Ca(2+) bursts and Crz1 activation in Saccharomyces cerevisiae. Microbiology **158**(Pt 9), 2293–2302 (2012)

22. K.J. Friston et al., Statistical parametric maps in functional imaging: a general linear approach. Hum. Brain Mapp. **2**(4), 189–210 (1995)
23. H. Ge et al., Comparative analyses of time-course gene expression profiles of the long-lived sch9Delta mutant. Nucleic Acids Res. **38**(1), 143–158 (2010)
24. C. González-Aguilera et al., Nab2 functions in the metabolism of RNA driven by polymerases II and III. Mol. Biol. Cell **22**(15), 2729–2740 (2011)
25. S. Gupta, J.A. Stamatoyannopolous, T. Bailey, W.S. Noble, Quantifying similarity between motifs. Genome Biol. **8** (2007)
26. J. Kleinberg, An impossibility theorem for clustering. Adv. Neural Inf. Process. Syst. 446–453 (2002)
27. S. Koelsch, Towards a neural basis of music-evoked emotions. Trends Cogn. Sci. **14**(3), 131–137 (2010)
28. S. Koelsch, Brain correlates of music-evoked emotions. Nat. Rev. Neurosci. **15**(3), 170–180 (2014)
29. S. Koelsch et al., Investigating emotion with music: an fMRI study. Hum. Brain Mapp. **27**(3), 239–250 (2006)
30. L.A.S. Kovacs et al., Cyclin-dependent kinases are regulators and effectors of oscillations driven by a transcription factor network. Mol. Cell **45**(5), 669–679 (2012)
31. A.M. Lanza, J.J. Blazeck, N.C. Crook, H.S. Alper, Linking yeast Gcn5p catalytic function and gene regulation using a quantitative, graded dominant mutant approach. PLoS ONE **7**(4), e36193 (2012)
32. M. Larsson et al., Functional studies of the yeast med5, med15 and med16 mediator tail subunits. PLoS ONE **8**(8), e73137 (2013)
33. M. Lee, C. Smyser, J. Shimony, Resting-state fMRI: a review of methods and clinical applications. Am. J. Neuroradiol. **34**, 1866–1872 (2013)
34. C. Liu, B. Abu-Jamous, E. Brattico, A.K. Nandi, Towards tunable consensus clustering for studying functional brain connectivity during affective processing. Int. J. Neural Syst. **27**(2), 1650042 (2017)
35. Z. Liu et al., Anaerobic α-amylase production and secretion with fumarate as the final electron acceptor in Saccharomyces cerevisiae. Appl. Environ. Microbiol. **79**(9), 2962–2967 (2013)
36. A.M. Matia-González, M.A. Rodríguez-Gabriel, Slt2 MAPK pathway is essential for cell integrity in the presence of arsenate. Yeast **28**(1), 9–17 (2011)
37. D.J. Montefusco et al., Distinct signaling roles of ceramide species in yeast revealed through systematic perturbation and systems biology analyses. Sci. Signal. **6**(299), rs14 (2013)
38. M. Monti, Statistical analysis of fMRI time-series: a critical review of the GLM approach. Front. Hum. Neurosci. **5**(609), 28 (2011)
39. M. Morillo-Huesca, M. Clemente-Ruiz, E. Andújar, F. Prado, The SWR1 histone replacement complex causes genetic instability and genome-wide transcription misregulation in the absence of H2A.Z. PLoS ONE **5**(8), e12143 (2010)
40. T. Nakamura et al., Identification of a gene, FMP21, whose expression levels are involved in thermotolerance in Saccharomyces cerevisiae. AMB Express **4**, 67 (2014)
41. D.A. Orlando et al., Global control of cell-cycle transcription by coupled CDK and network oscillators. Nature **453**, 944–947 (2008)
42. L.S. Parreiras, L.M. Kohn, J.B. Anderson, Cellular effects and epistasis among three determinants of adaptation in experimental populations of Saccharomyces cerevisiae. Eukaryot. Cell **10**(10), 1348–1356 (2011)
43. E.D. Raj, L.D. Babu, An enhanced trust prediction strategy for online social networks using probabilistic reputation features. Neurocomputing **219**, 412–421 (2017)
44. S. Saarikallio et al., Dance moves reflect current affective state illustrative of approach–avoidance motivation. Psychol. Aesthet. Creat. Arts **7**(3), 296–305 (2013)
45. V. Salimpoor, R. Zatorre, Neural interactions that give rise to musical pleasure. Psychol. Aesth. Creat. Arts **7**(1), 62–75 (2013)

46. A.B. Sanz et al., Chromatin remodeling by the SWI/SNF complex is essential for transcription mediated by the yeast cell wall integrity MAPK pathway. Mol. Biol. Cell **23**(14), 2805–2817 (2012)
47. SGD, Term Finder tool (2014). http://www.yeastgenome.org/cgi-bin/GO/goTermFinder.pl
48. K. Strassburg et al., Dynamic transcriptional and metabolic responses in yeast adapting to temperature stress. OMICS **14**(3), 249–259 (2010)
49. A. Strehl, J. Ghosh, Cluster ensembles—a knowledge reuse framework for combining multiple partitions. J. Mach. Learn. Res. **3**, 583–617 (2002)
50. T. Suzuki, Y. Iwahashi, Gene expression profiles of yeast Saccharomyces cerevisiae sod1 caused by patulin toxicity and evaluation of recovery potential of ascorbic acid. J. Agric. Food Chem. **59**(13), 7145–7154 (2011)
51. T. Suzuki, Y. Iwahashi, Comprehensive gene expression analysis of type B trichothecenes. J. Agric. Food Chem. **60**(37), 9519–9527 (2012)
52. B. Thirion, G. Varoquaux, E. Dohmatob, J.-B. Poline, Which fMRI clustering gives good brain parcellations? Front. Neurosci. **8**, 1–13 (2014)
53. D. Tomasi, N. Volkow, Abnormal functional connectivity in children with attention-deficit/hyperactivity disorder. Biol. Psychiatry **71**(5), 443–450 (2012)
54. N. Tzourio-Mazoyer et al., Automated anatomical labeling of activations in SPM using a macroscopic anatomical parcellation of the MNI MRI single-subject brain. NeuroImage **15**(1), 273–289 (2002)
55. M. van den Heuvel, H. Pol, Exploring the brain network: a review on resting-state fMRI functional connectivity. Eur. Neuropsychopharmacol. **20**(8), 519–534 (2010)
56. S.L. Wade, K. Poorey, S. Bekiranov, D.T. Auble, The Snf1 kinase and proteasome-associated Rad23 regulate UV-responsive gene expression. EMBO J. **28**(19), 2919–2931 (2009)
57. R.W. Wilkins et al., Network science and the effects of music preference on functional brain connectivity: from Beethoven to Eminem. Sci. Rep. **4**, 6130 (2014)
58. Y. Xue-Franzén, J. Henriksson, T.R. Bürglin, A.P. Wright, Distinct roles of the Gcn5 histone acetyltransferase revealed during transient stress-induced reprogramming of the genome. BMC Genom. **14**, 479 (2013)

Chapter 3
Do You Mean What You Say? Recognizing Emotions in Spontaneous Speech

Rupayan Chakraborty, Meghna Pandharipande and Sunil Kumar Kopparapu

Abstract Emotions when explicitly demonstrated by an actor are easy for a machine to recognize by analyzing their speech. However in case of day to day, naturally spoken spontaneous speech it is not easy for machines to identify the expressed emotion even though emotion of the speaker are embedded in their speech. One of the main reasons for this is that people, especially non-actors, do not explicitly demonstrate their emotion when they speak, thus making it difficult to recognize the emotion embedded in their spoken speech. In this paper, based on some of our previous published work (example, Chakraborty et al. in Proceedings of the 20th International Conference KES-2016 96:587–596, 2016 [1], Chakraborty et al. in TENCON 2015—2015 IEEE Region 10 Conference 1–5, 2015 [2], Chakraborty et al. in PACLIC, 2016 [3], Pandharipande and Kopparapu in TENCON 2015—2015 IEEE Region 10 Conference 1–4, 2015 [4], Kopparapu in Non-Linguistic Analysis of Call Center Conversations, 2014 [5], Pandharipande and Kopparapu in ECTI Trans Comput Inf Technol 7(2):146–155, 2013 [6], Chakraborty and Kopparapu in 2016 IEEE International Conference on Multimedia and Expo Workshops, 1–6, 2016 [7]) we identify the challenges in recognizing emotions in spontaneous speech and suggest a framework that can assist in determining the emotions expressed in spontaneous speech.

Keywords Spontaneous speech · Emotion · Recognition · Classification · Challenges

R. Chakraborty · M. Pandharipande · S.K. Kopparapu (✉)
TCS Innovation Labs, Mumbai, India
e-mail: SunilKumar.Kopparapu@TCS.Com

R. Chakraborty
e-mail: Rupayan.Chakraborty@TCS.Com

M. Pandharipande
e-mail: Meghna.Pandharipande@TCS.Com

S.R.S. Prabaharan et al. (eds.), *Frontiers in Electronic Technologies*,
Lecture Notes in Electrical Engineering 433, DOI 10.1007/978-981-10-4235-5_3

1 Introduction

Several nuances are embedded in human speech. A spoken utterance can be analyzed for *what* was spoken (speech recognition), *how* was it spoken (emotion recognition) and *who* spoke (speech biometric) it. Most often these three aspects form the basis of most of the work being carried out actively by speech researchers. In this paper, we concentrate on the *how* aspect of spoken utterance, namely emotion recognition.

Perceiving emotions from different real-life signals is a natural and an inherent characteristic of a human being. For this reason emotion plays a very important role in intelligent human–computer interactions. Machine perception of human emotion not only helps machine to communicate more humanely, but it also helps in improving the performance of other associated technologies like Automatic Speech Recognition (ASR) and Speaker Identification (SI).

With the mushrooming of services industry there has been a significant growth in the voice-based call centers (VbCC) where identifying emotion in spoken speech has gained importance. The primary goal of a VbCC is to maintain a high level of customer satisfaction which means understanding the customer just in time (in real time and automatically) and making a decision on how to communicate (what to say, how to say) with the customer. While several things related to the customer are a priori available, thanks to the advances in data mining, the one thing that is crucial is the emotion of the customer at that point of time, so that the agent can plan what and how to converse to keep the customer happy and also allow him to know when to make a pitch to up-sell.

Much of the initial emotion recognition research has been successfully validated on acted speech (for example [8–11]). Emotions expressed by trained actors are easy to recognize, primarily because they are explicitly and dramatically expressed by them with significant intensity. This, on purpose magnified, emotions can be easily distinguished from one another. However, when the expression of the emotion is not explicit or loud, it is very difficult to distinguish one emotion of the speaker from another. This mild and not explicitly demonstrated emotion is most likely to occur in spontaneous natural day to day conversational speech. The rest of the paper is organized as follows. In Sect. 2 we dwell on the different challenges facing emotion recognition in spontaneous speech. We propose a framework in Sect. 3 that has provision to use prior knowledge to address emotion recognition in spontaneous speech. And we conclude in Sect. 4.

2 Challenges

Historically emotion has been represented using two affective dimensions, namely, *arousal* (also referred to as activation) and *valence*. Note that any point in this 2D space (Fig. 1) can be looked upon as a vector and represents an emotion. Table 1 gives the mapping of a few emotions in terms of the affective states. For

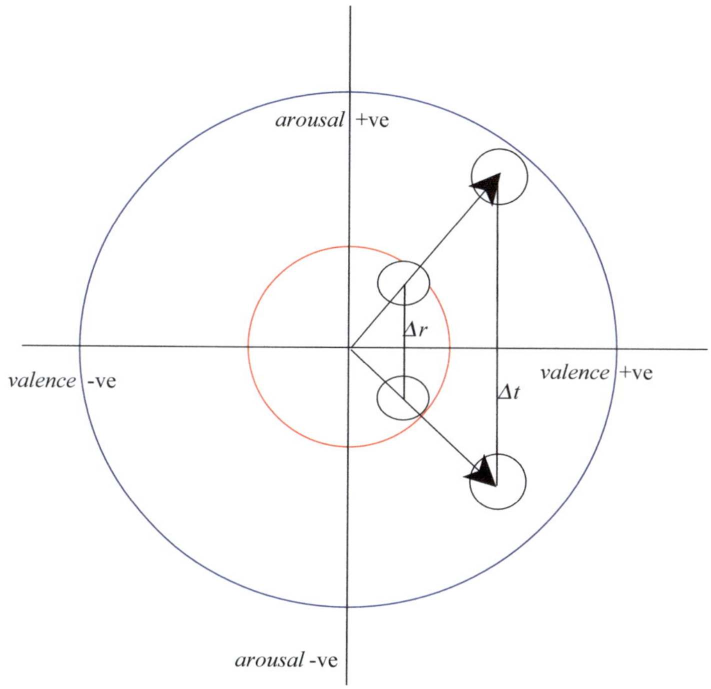

Fig. 1 Emotions expressed in the (*arousal*, *valence*) space. Emotion in spontaneous speech is subtle compared to acted speech

Table 1 Emotions expressed in the (*arousal*, *valence*) space. Map of the affective states to known human emotion

arousal	*valence*	Emotion
+	+	*happy*
+	−	*anger*
0	0	*neutral*
−	−	*sad*

example +ve *valence* and +ve *arousal* (first quadrant) would represent *happy* while −ve *valence* and +ve *arousal* could represent *anger*. Now we enumerate the challenges in machine recognizing emotion in spontaneous speech.

Intensity of Emotion in Spontaneous Speech Usually acted speech exhibits higher degree of intensity, both in *arousal* and *valence* dimensions resulting in a larger radii emotion vector compared to the spontaneous (non-acted) speech. For this reason, it is easy to mis-recognize one emotion for another in spontaneous speech. Subsequently, if the first quadrant (Fig. 1) represents emotion E_1 and the fourth quadrant represents emotion E_2, then the misrecognition error is small (Δr) for spontaneous speech but requires higher degree of error in judgment (Δt) to mis-recognize emotion E_1 as emotion E_2 and vice-versa for acted speech. For this reason, recognizing emotion in spontaneous speech becomes challenging and is more prone to misrecognition.

What works for acted speech does not work for Spontaneous Speech Though recognizing emotion in speech has a rich literature, however, most of the work has been done on acted speech are typically machine learning-based systems. Namely, one trains a system (example, Support Vector Machine, Artificial Neural Networks, Deep Neural Networks) with a set of annotated speech data and then classifies the test dataset using the trained system. Speech emotion recognition systems that perform with high accuracies on acted speech datasets do not perform as well on realistic natural speech [12] because of the mismatch between the train (acted) and test (spontaneous) datasets. This is another challenge in addressing spontaneous speech emotion recognition.

Clearly this challenge can be addressed if there exists an emotion annotated spontaneous speech dataset which can be used for training a spontaneous speech emotion recognition system.

Spontaneous Speech Corpus For any given spoken utterance, there are two point of views in terms of associating emotion to the utterance namely, (a) encoded emotion and (b) decoded emotion. The emotional state of the speaker who uttered the audio is called the encoded emotion while the interpreted emotion of the same audio by a listener, who is different from the speaker, is called decoded emotion. For example, the audio $s(t)$ in Fig. 2 can have two emotion labels associated with it. When the speaker annotates and assigns a emotion label it is called the encoded emotion of $s(t)$ and when a listener (different from the speaker) listens to $s(t)$ and assigns an emotion label it is called the decoded emotion.

For acted speech both decoded and encoded emotion are more likely to be the same, however, for spontaneous speech there is bound to be a wide gap between the encoded and decoded emotion. Building a realistic spontaneous speech corpus would need a person to speak in a certain emotional state and/or annotate what

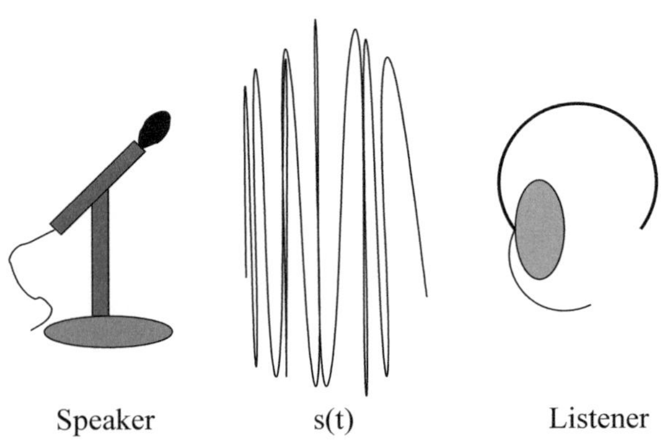

Fig. 2 Decoded versus encoded emotions

he or she spoke; generating such realistic spontaneous data corpus is extremely difficult and is a huge challenge.

Annotating Spontaneous Speech Corpus The next best thing is to have a decoded spontaneous speech corpus. However, one of the problems associated with emotion recognition of spontaneous speech, is the availability of a reliably emotion annotated spontaneous speech database suitable for emotion recognition. The inability to annotate spontaneous speech corpus is basically because of the lower degree of emotion expression (as seen in Fig. 1).

In [1] we showed that there is a fair amount of disagreement among the evaluators when they are asked to annotate spontaneous spoken utterances. The disagreement, however, decreases when the evaluators are provided with the context knowledge associated with the utterance. Fleiss' Kappa score [13, 14] was used to determine the agreement between evaluators. When the evaluators were asked to annotate (decoded emotion) spontaneous speech the agreement was 0.12 while the same set of evaluators when provided with the context associated with the spontaneous speech, the agreement between the evaluators increased to 0.65. This suggests that there is a higher degree of agreement between the evaluators when they are provided associated contextual knowledge while annotating spontaneous speech.

As illustrated above, there are several known challenges that exist in spontaneous speech emotion recognition. Clearly the literature that deals with emotion recognition of acted speech does not help in spontaneous speech emotion recognition, however, as observed, the use of prior knowledge can help address recognizing emotions in spontaneous speech. In the next section we propose a framework for recognizing emotions in spontaneous speech based on this observation.

3 A Framework for Spontaneous Speech Emotion Recognition

Let $s(t)$ be a speech signal, say of duration T seconds and let

$$\mathscr{E} = (E_1 = anger, E_2 = happy, \ldots, E_n)$$

be the set of n emotion labels. In literature the emotion of the speech signal $s(t)$ is computed as

$$\mu^p_{k,s(t)} = P(E_k|s(t)) = \frac{P(s(t)|E_k)P(E_k)}{P(s(t))} \tag{1}$$

where $\mu^p_{k,s(t)} = P(E_k|s(t))$ is the posterior probability or score associated with $s(t)$ being labeled as emotion $E_k \in \mathscr{E}$. Generally, these posteriors are calculated by learning the likelihood probablities from a reliable training dataset using some machine learning algorithm. Note that in practice the features extracted from the speech signal

$\mathscr{F}(s(t))$ are used instead of the actual raw speech signal $s(t)$ in (1). Conventionally, the emotion of the speech signal $s(t)$ is given by

$$E_{k*} = \arg \max_{1 \le k \le n} \{\mu^{p}_{k,s(t)}\}. \tag{2}$$

Note that $E_{k*} \in \mathscr{E}$ is the estimated emotion of the signal $s(t)$.

While this process of emotion extraction works well for acted speech, because the entire speech utterance carries one emotion. However this, namely the complete speech signal carrying a single emotion is seldom true for spontaneous conversational speech (for example, a call center audio recording between the agent and a customer). As mentioned in an earlier section, additional challenges exists in terms of the fact that emotions in spontaneous speech are not explicitly demonstrated and hence can not be robustly identified even by human annotators in the absence of sufficient context surrounding the spoken utterance.

These observations lead us to look for a novel framework for recognizing emotions in spontaneous speech [1]. The framework tries to take care of the fact that (a) the emotion within the same speech signal is not the same and (b) human annotators are better able to recognize emotions when they are provided with a context associated with the speech signal.

The essential idea of the framework is to compute emotion for smaller duration $(2\Delta\tau)$ segments of the speech signal $(s_\tau(t))$, namely,

$$s_\tau(t) = s(t) \times \{U(\tau - \Delta\tau) - U(\tau + \Delta\tau)\}$$

where $U(t)$ is a unit step function defined as

$$\begin{aligned} U(t) &= 1 \quad \text{for} \quad t \ge 0 \\ &= 0 \quad \text{for} \quad t < 0, \end{aligned}$$

instead of computing the emotion for the complete signal $(s(t))$. Note that (a) $s_\tau(t) \subset s(t)$ and is of length $2\Delta\tau$ and (b) $\tau \in [0, T]$. As done conventionally, the emotion of $s_\tau(t)$ is computed as earlier, namely,

$$\mu^{p}_{k,s_\tau(t)} = P(E_k | s_\tau(t)) \tag{3}$$

for $k = 1, 2, \ldots n$. However, in addition we also make use of the emotions computed from the previous η speech segments, namely $\mu_{k,s_{\tau-v}(t)}$ for $v = 1, 2, \ldots \eta$. So we have, the posterior score associated with the speech utterance $s_\tau(t)$ being labeled E_k as

$$'\mu^{p}_{k,s_\tau(t)} = \mu^{p}_{k,s_\tau(t)} + \sum_{v=1}^{\eta} \omega_v \mu^{p}_{k,s_{\tau-v}(t)} \tag{4}$$

where $\omega_1, \omega_2 \ldots, \omega_\eta$ are monotonically decreasing weights, which are all less than 1. Equation (4) makes sure that the posterior score of the speech segment $s_\tau(t)$ is influ-

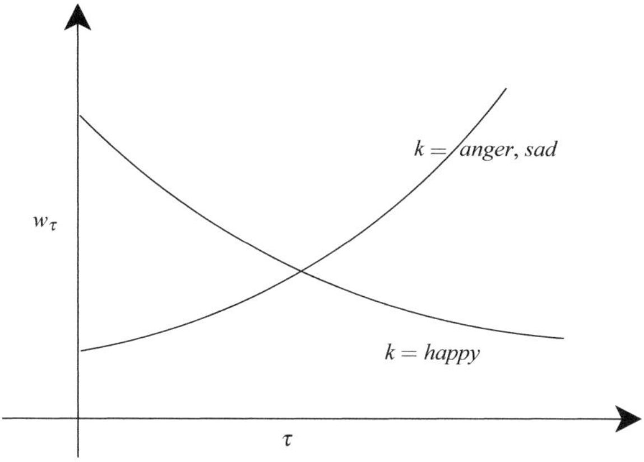

Fig. 3 Knowledge regarding the time lapse of the utterances in the call. The weights w_τ of emotions like *happy* decreases with τ while the weights increase for emotions like *anger, sad*

enced by the weighted sum of the posterior score of the previous speech segments. This is generally true of spontaneous conversational speech where the emotion of the speaker is based on the past emotion experienced during the conversation.

Further the output posterior scores from emotion recognizer, namely, $\mu^{p}_{k,s_\tau(t)}$ (3) is given as input to a knowledge-based system, that modifies the scores depending upon the time lapse (how far is τ from the beginning of the spoken conversation) of the speech segment (utterance) in the audio signal. This can be represented as,

$$\mu^{\kappa}_{k,s_\tau(t)} = w_\tau \mu^{p}_{k,s_\tau(t)} \tag{5}$$

where $\mu^{p}_{k,s_\tau(t)}$ (3) and w_τ (see Fig. 3) are the posterior probability score and weight vector at time instant τ respectively. And $\mu^{\kappa}_{k,s_\tau(t)}$ is the emotion computed based on knowledge.

The motivation for (5) is based on the observation that the duration of the audio calls plays an important role in the induction (or change) in the user's emotion. As mentioned in [1] weight w_τ is expected to increase or decay exponentially as τ increases, depending upon the type of the emotion. As an example (see Fig. 3) it is expected that w_τ for *anger* and *sad* close to the end of the conversation is likely to be more compared to the same emotion of the customer at the beginning of the call. As seen in Fig. 3 the weight components are expected to increase exponentially as time index increases for *anger* and *sad* at the same time w_τ is expected to decrease exponentially as time index increases for *happy* emotion.

We can combine μ^{p}_{k} and μ^{κ}_{k} to *better* estimate the emotion of the spontaneous utterance $s_\tau(\tau)$ as

$$e^{k} = \lambda_{p}('\mu^{p}_{k}) + \lambda_{\kappa}(\mu^{\kappa}_{k}) \tag{6}$$

where $\lambda_\kappa = 1 - \lambda_p$. The framework makes use of knowledge when $\lambda_\kappa \neq 0$. Emotion of the spontaneous speech utterance $s(\tau)$ with the incorporation of knowledge (μ^{κ}_{k}) is represented as

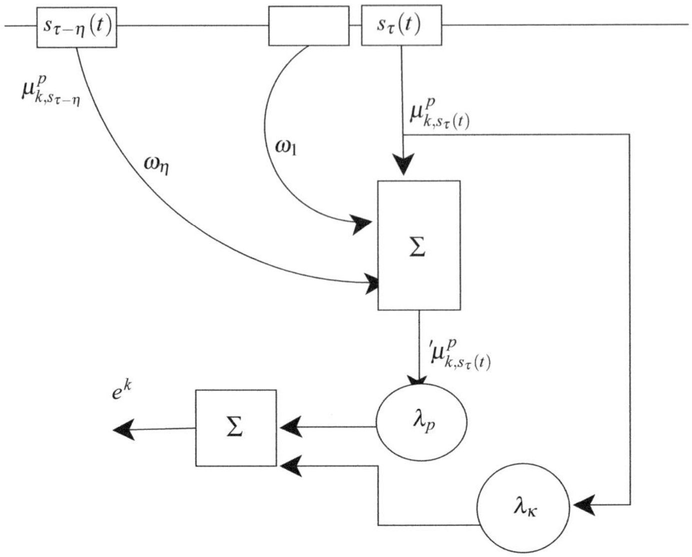

Fig. 4 Proposed framework for spontaneous speech emotion recognition

$$E_{k*} = \arg \max_{1 \le k \le n} \left\{ e^k \right\}. \quad (7)$$

Knowledge regarding the time lapse of the utterance in an audio call, especially in conversational system, provides useful information to recognize the emotion of the speaker. Therefore, incorporation of this knowledge is useful in extracting the actual emotion of an user. The proposed framework for spontaneous speech recognition is shown in Fig. 4.

As shown in [1] there is performance improvement in recognition of emotion of spontaneous speech when this framework is used. They show for different classifiers that there is almost 11% absolute improvement in emotion recognition for interactive voice response type of call and the performance further improved to 14% absolute for real call center conversation.

4 Conclusion

Emotion recognition has rich literature for acted speech and this leads to the belief that the techniques that work well for acted speech can be directly used for spontaneous speech. However, there are several dissimilarities between acted and spontaneous speech which does not allow one to use techniques and algorithms that work well for acted speech to recognize emotion in spontaneous speech. Emotion recognition techniques are generally machine learning based algorithms which (a) requires sufficient amount of training data and (b) requires the test and the train data to match. The main challenge in using trained models that work for acted speech on spontaneous speech is the mismatched condition. Additionally, in case of spontaneous speech it is very challenging to (a) generate spontaneous speech data and (b) to obtain robust annotation of the speech data. For this reason techniques and algorithms that work best for spontaneous speech cannot be built afresh. In this paper, we first established the importance of spontaneous speech emotion recognition and then enumerated several challenge and hurdles faced during emotion recognition in spontaneous speech. Based on our previous work, we proposed a framework that exploits apriori knowledge to enable reliable spontaneous speech emotion recognition. The main idea behind the proposed framework is to assist the machine learning algorithm with prior knowledge associated with the spontaneous speech. It has been shown [1] that this framework can actually improve the emotion recognition accuracies of spontaneous speech by as much as 11–14% in absolute terms.

References

1. R. Chakraborty, M. Pandharipande, S.K. Kopparapu, Knowledge-based framework for intelligent emotion recognition in spontaneous speech, in *Procedia Computer Science, 2016, knowledge-Based and Intelligent Information; Engineering Systems: Proceedings of the 20th International Conference KES-2016*, vol. 96, pp. 587–596. http://www.sciencedirect.com/science/article/pii/S187705091632049X
2. R. Chakraborty, M. Pandharipande, S. Kopparapu, Event based emotion recognition for realistic non-acted speech, in *TENCON 2015—2015 IEEE Region 10 Conference* (2015), pp. 1–5
3. R. Chakraborty, M. Pandharipande, S.K. Kopparapu, Mining call center conversations exhibiting similar affective states, in *PACLIC 2016* (2016)
4. M.A. Pandharipande, S.K. Kopparapu, Audio segmentation based approach for improved emotion recognition, in *TENCON 2015—2015 IEEE Region 10 Conference* (2015), pp. 1–4
5. S.K. Kopparapu, *Non-Linguistic Analysis of Call Center Conversations*, Springer Briefs in Electrical and Computer Engineering (Springer, 2014)
6. M.A. Pandharipande, S.K. Kopparapu, A language independent approach to identify problematic conversations in call centers. ECTI Trans. Comput. Inf. Technol. **7**(2), 146–155 (2013)
7. R. Chakraborty, S.K. Kopparapu, Improved speech emotion recognition using error correcting codes, in *2016 IEEE International Conference on Multimedia and Expo Workshops, ICME Workshops 2016, Seattle, WA, USA, July 11–15, 2016*. IEEE Computer Society (2016), pp. 1–6. doi:10.1109/ICMEW.2016.7574707
8. B.W. Schuller, A. Batliner, S. Steidl, D. Seppi, Recognising realistic emotions and affect in speech: state of the art and lessons learnt from the first challenge. Speech Commun. **53**, 1062–1087 (2011)
9. M.E. Ayadi, M.S. Kamel, F. Karray, Survey on speech emotion recognition: features, classification schemes, and databases. Pattern Recogn. **44**, 572–587 (2011)
10. E. Mower, M. Mataric, S.S. Narayanan, A framework for automatic human emotion classification using emotion profiles. IEEE TASLP **19**(5), 1057–1070 (2011)
11. S. Wu, T.H. Falk, W.Y. Chan, Automatic speech emotion recognition using modulation spectral features. Speech Commun. **53**(5), 768–785 (2010)
12. B.W. Schuller, D. Seppi, A. Batliner, A.K. Maier, S. Steidl, Towards more reality in the recognition of emotional speech, in *ICASSP* (2007), pp. 941–944
13. A.J. Viera, J.M. Garrett, Understanding interobserver agreement: the kappa statistic. Family Med. **37**(5), 360–363 (2005)
14. M.L. McHugh, Interrater reliability: the kappa statistic. Biochemia Med. **3**, 276–282 (2012)

Chapter 4
Sensors and Actuators for HCI and VR: A Few Case Studies

Daniel Thalmann

Abstract This paper discusses the role of sensors and actuators in Human–Computer Interaction and Virtual Reality. We first introduce the need of sensory input and output and the user senses stimulation. We then present two experiences of immersive games with a comparison of the techniques 20 years ago and today and also discuss the importance of vision-based hand tracking. The main part of the paper is dedicated to haptic and tactile feedbacks with 4 different case studies: haptic feedback using a haptic workstation, a wearable system for mobility improvement of visually impaired people, a system for enhancing pilot performance, and actuators to feel the wind. Finally, we survey a few smart jackets with sensors and actuators.

Keywords Virtual reality · Sensors · Actuators · Haptics · Immersion · Hand tracking · Smart jacket

1 Introduction

Since many years, we hear about Virtual Reality as a discipline that could potentially provide benefits to many applications. Basically, the principle of Virtual Reality consists in stimulating user senses in order to give the impression to be in another place which they can discover and interact with. Already in 1965, Ivan Sutherland defined the "Ultimate Display" [1]: "It consists in a room within which the computer can control the existence of matter. A chair displayed in such room would be good enough to sit in." This classical paper introduced the key concept of complete sensory input and output using specific hardware capable of stimulating

D. Thalmann (✉)
Institute for Media Innovation, Nanyang Technological University, Singapore, Singapore
e-mail: danielthalmann@ntu.edu.sg
URL: https://en.wikipedia.org/wiki/Daniel_Thalmann

D. Thalmann
EPFL, Lausanne, Switzerland

S.R.S. Prabaharan et al. (eds.), *Frontiers in Electronic Technologies*, Lecture Notes in Electrical Engineering 433, DOI 10.1007/978-981-10-4235-5_4

human senses in a similar way than in reality. Today, most virtual reality systems create realistic visual and auditive environments. Nowadays, Ivan Sutherland is able to see the virtual chair he was dreaming about, but he will still experience some problems if he tries to sit on it. However, the interaction with these environments cannot be considered as natural. Indeed, we often use devices such as mouse, keyboard, or joystick to move or manipulate them. These paradigms of interaction are in fact metaphors, which are not similar to reality. In some situations they are practical and efficient. However, the lack of intuitiveness sometimes makes them limited or simply ineffective.

Virtual Reality in its most "classical" form requires real-time graphics, a stereoscopic display, used to produce the illusion of 3D, and a tracking system to acquire head and hands motion. Commonly used technologies include head-mounted displays (HMD) and stereoscopic glasses. Interaction can be achieved using a tracking device, which may be integrated in the HMD itself, for tracking head and body movement. A "data glove" can be used to track hand movements.

VR is about simulating reality. The aspect of reality that has been most frequently addressed is the visual one. Sight is, for most people, the dominant perceptual sense and the principal means for acquiring information. However, reality is not limited to what we can see. Other important components of our perceptual experience are sounds and tactile feedback.

Developing VR systems involves different disciplines that address each of the human senses: computer graphics (sight), 3D sound synthesis (hearing), and haptics (touch). Smell and taste play important roles in our daily life, but have been less exploited in VR due to the complexity of the required technology. Recently, according to Business Insider, Oculus Rift chief technical officer Mike Schroepfer just told reporters at the Dublin Web Summit that he wants the Oculus Rift to become "a device that allows you to be anywhere you want, with anyone, regardless of geographic boundaries" [2].

To overcome this, researchers can use Haptic Devices. They are designed to simulate what is commonly called the "sense of touch," which includes more specifically, tactile, pain sense, thermal sense, and proprioception. Proprioception is knowledge gained by the perception of the relative member's position of the human body. There are two advantages of such haptic devices. First, they can give the user more information on the nature of virtual objects (size, weight, finish, rigidity, etc.). Second, they can provide interaction paradigms that are closer to reality (three-dimensional interaction in a three-dimensional world). However, haptic device mechanics is complex.

Palmer Luckey admitted it would be "a long time" before you will even be able to feel a fuzzy couch in VR. Both Luckey and Oculus chief scientist Michael Abrash have consistently reminded people that we are at the very beginning of virtual reality; they think in half-centuries, not years or even decades.

Unfortunately, a surprising amount of VR rhetoric elides the huge problem that is touch, treating it like a nice extra feature instead of a key to the whole experience. We usually touch object using our hands and finger tip. The other area of our body's skin is still capable of feeling contact but we usually do not use this sensory capability. Most of the time we just keep this sensation to feel comfortable by wearing soft clothing and keep us at a comfortable temperature. In those areas our tactile sense is limited when it comes to its use to gather information. We usually keep distance with other persons, the tactile contact between two individual is something we keep very intimate.

In terms of research, the key question is what kind of information is suitable to be passed through tactile stimulation of the body's skin. It leads to further problematic, in which extend using the tactile sense of our body's skin can replace other missing senses, give information about distant environment, or enhance physical resistance to perform task under difficult constraints.

Proprioception and mechanical stimulation at the skin level are link to motor reflexes. The whole sensory motor system responsible for postural sway is affected by vibrotactile stimulation when visual and vestibular cues are absent or disturbed. Those experiments put in evidence the fact that tactile stimulation could not only affects our most primitive reflexes but can alter the perception of our own body posture in space. Using this property, one interesting aspect is to study response time to diverse stimulation and compare them to other cues such as sight and hearing.

Touch is a proximal sense, we feel things close to us or actually in contact with us. Although some exceptions occur, for example heat radiation and deep bass tones. Another interesting aspect is that it can be extended using simple tools like a long cane provides vibratory and pressure information for a blind. The sense of touch is synonym to tactual perception and Touch often considered one of five human senses defined by Aristote. However, the sense of touch evokes various feelings from pressure to temperature and pain. Thus the term "touch" is actually a combined term for several sub-modalities, in medicine touch is usually replaced with the term somatic senses to better reflect the variety of sensory mechanisms involved. The senses of touch are mediated by the somatosensory system.

As mentioned by Pakala et al. [3], sensors and actuators are critical components of an Embedded Systems (ES) in a large number of complex real life applications, for example, Radars, Aircraft, Process Industry, and host of other systems that makes use of automatic control.

Transducers are sensors and actuators that allow a technical system to interact with the physical environment. An intelligent or "smart" transducer is the integration of an analog or digital sensor or actuator element, a processing unit, and a communication interface.

According to analyst IHS, by 2019, the wearables market is estimated to reach more than 230 million units, comprised of various devices promising always-on health monitoring, leisure connectivity, environment automation, and a broad range of security functions. The trend to incorporate more sensors into wearable devices will accelerate in the coming years, giving us not only a wider spread of information about the world around us but also making it possible to understand higher level patterns in our activities and helping us to make better lifestyle choices.

But what about actuators?

The Mexican company Vivoxie has created Power Claw [4], a pair of gloves with an interface that stimulates the skin and allows the sense of touch in cyber worlds. The device generates sensations of heat, cold, vibration, and roughness of objects in a virtual reality simulation. The gloves are complemented with Oculus Rift glasses.

2 Playing Games Together

2.1 Immersive Games

VR is the ideal technology to provide the way of playing together to collective games. One of the first game developed is a tennis game proposed by Molet et al. [5] At the opening and closing session of Telecom Interactive '97 in Geneva, Switzerland, the authors presented in real time a virtual, networked, interactive tennis game simulation (see Fig. 1). In this demonstration the interactive players were merged into the virtual environment by head-mounted displays, magnetic sensors and data gloves. There were two players: the University of Geneva player was "live" on stage at the opening session and the other player in EPFL at Lausanne, separated by a distance of approximately 60 km. For managing and controlling the shared networked virtual environment we used the Virtual Life Network (VLnet), which is a general purpose client/server network system using realistic virtual humans (avatars) for user representation. These avatars support body deformation during motion. The virtual humans also represent autonomous virtual actors such as the synthetic referee, who is part of the tennis game simulation. A special tennis ball driver animated the virtual ball by detecting and treating collisions between the tennis ball, the virtual rackets, the court, and the net.

About 20 years later, we developed a volleyball game allowing to play not only against opponents, but also in cooperation with teammates. Volleyball is a team sport in which two teams of six players are separated by a net. Each team tries to score points by grounding a ball on the other team's court following rules. Our

Fig. 1 Anyone for tennis using VLnet

system provides immersive volley ball game with interactive virtual players. In this immersive volley ball game, a user will be located in the center area as shown in Fig. 2a. Then, s/he will collaborate with five virtual teammates as shown in Fig. 2b. The six virtual opponents will play against user's team as shown in Fig. 2c. The immersive volleyball game can run on various 3D displays using the same input interfaces. The user can be immersed and interact in the game with the two hands. To do this, the user wears head tracking markers on his/her head, and he/she uses two types of infrared markers for the tracking of the hand 3D positions. The Sensor Manager receives the inputs from the user in real time and then, handles and delivers the inputs to the Coordination Manager. In the Coordination Manager, both head and hand tracking information are transformed into unified coordinates. The converted data are provided to the Game Manager in which, the viewpoint is updated by the head tracked position, and the virtual two hands are mapped to the

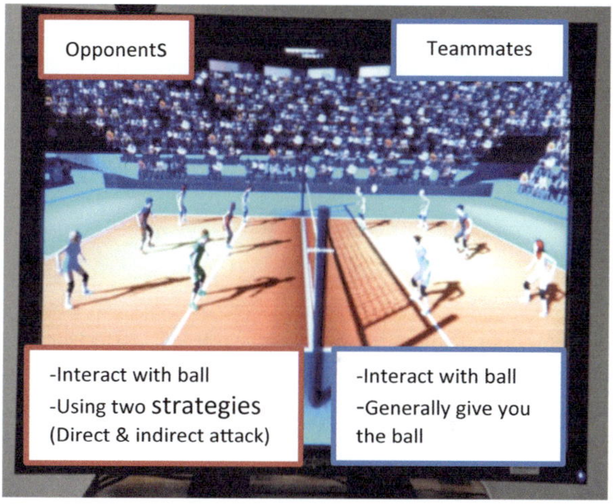

Fig. 2 Volley ball game. **a** User area. **b** Teammates area. **c** Opponents area

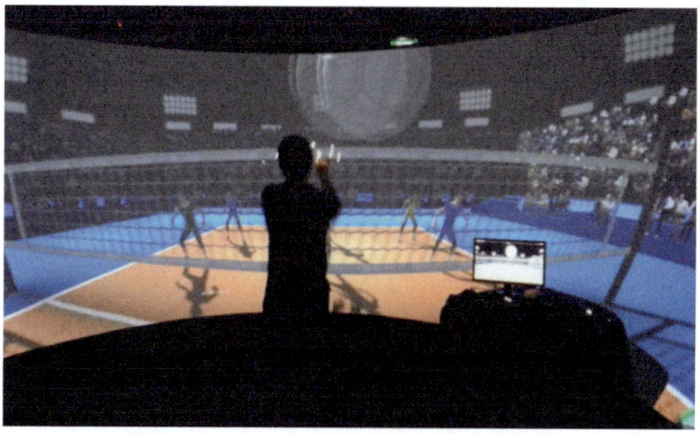

Fig. 3 The game in the immersive room

Fig. 4 The game through the oculus rift

hand tracked information. The Virtual Human Manager manages 3D virtual humans and behaviors. Finally, the results are visualized to the different 3D displays as the IMI Immersive Room (Fig. 3) and the Oculus Rift (Fig. 4).

2.2 *Vision-Based Hand Tracking*

Hand pose estimation is an important research topic in human–computer interaction that has various applications, such as gesture recognition and animation synthesis and is a key issue for immersive gaming. Previously the specialized hardware, e.g., the optical sensors [6] and the data gloves [7], were commonly used to accomplish this task. Although they provide accurate measurements and achieve real-time performance, such devices are cumbersome to use and expensive. Thus, the vision-based methods have been the mainstream in this field, which are cheaper and provide more natural interaction experiences. However, due to the high flexibility and self-occlusion of the hand, it remains a challenging task to capture the articulated hand motions from the visual inputs.

The purpose of full-DOF hand pose estimation is to recover the 3D positions of a set of hand joints. In our previous work we have tested and developed various techniques for full-DOF hand pose estimation, and find out that the random regression forest-based methods can achieve reasonably high accuracy at low computational complexity. However, regression forest usually suffers from ambiguous predictions. Therefore, we developed a system to estimate the articulated hand pose with an emphasis on reducing prediction ambiguity by exploiting the hand part correlations [8]. Figure 5 shows the steps of the method.

More recently, Ge et al. [9] propose a novel 3D regression method using multi-view CNNs which can better exploit depth cues to recover fully 3D information of hand joints without model fitting. Figure 6 illustrates the principle.

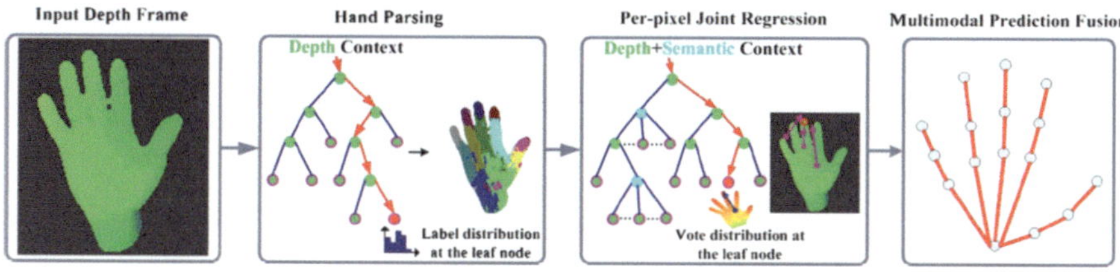

Fig. 5 Full-DOF hand pose estimation using per-pixel regression and multimodal prediction fusion

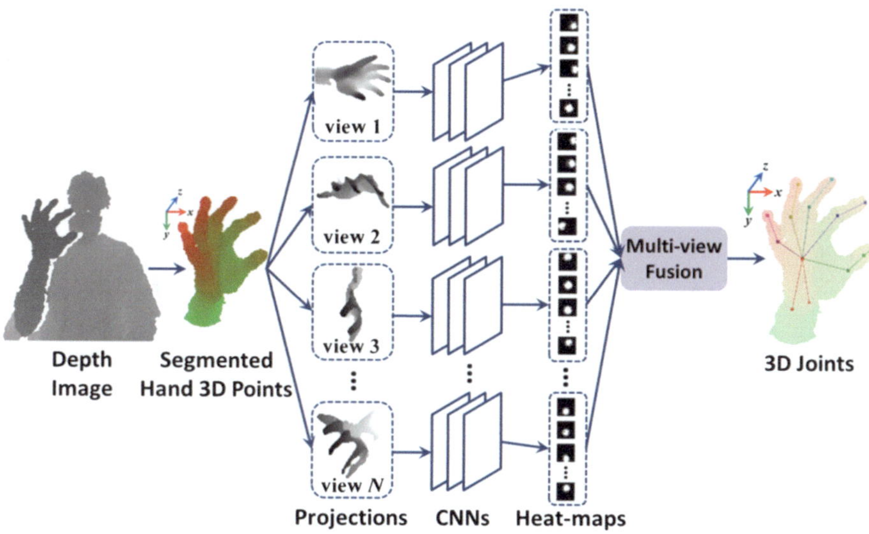

Fig. 6 Full-DOF hand pose estimation using multi-view fusion

3 Haptic and Tactile Actuators

3.1 Haptic Feedback

Our main experience [10] is based on the Immersion Haptic Workstation as presented in Fig. 7. The organization of the software system is shown in Fig. 8. Our haptic feedback system is based on a proxy approach [11]. A proxy is weakly linked to the position of the device, i.e., the Haptic Workstation. Basically, it is based on three hand models (see Fig. 9):

- The Tracked Hand, which is the virtual hand skeleton created after calibration. It almost corresponds to the representation of the real hand position orientation and posture into the Virtual Environment; it is not exact, but we consider that the matching is correct.

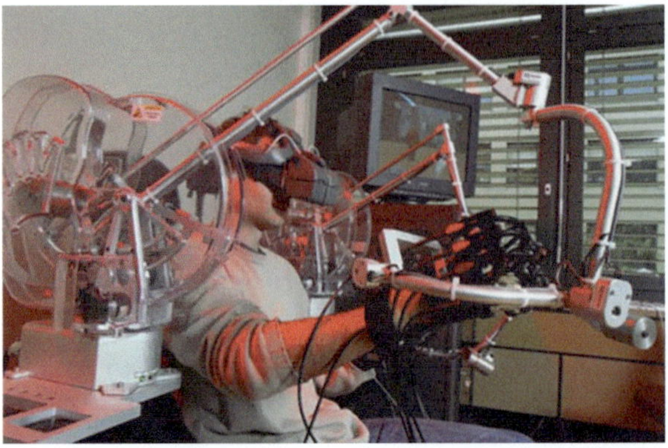

Fig. 7 The haptic workstation (EPFL)

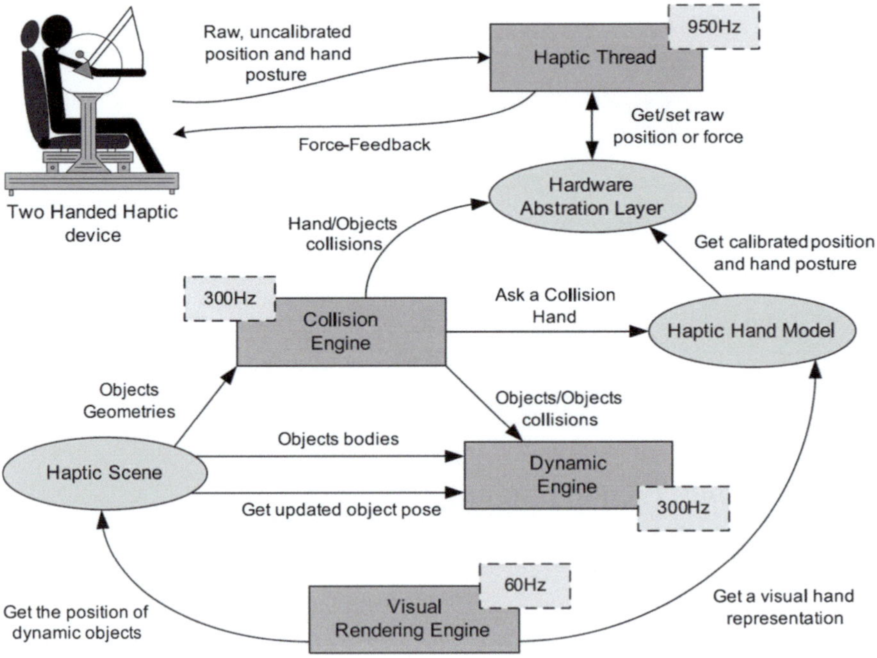

Fig. 8 Organization of the haptic feedback system in EPFL

- The Proxy hand, which is a mass-spring-damper system with the shape of the hands. Each phalanx and the palm have a special geometry for collision processing with also dynamic properties. These elements are linked together with motorized joints parameterized with springs and damping coefficient.

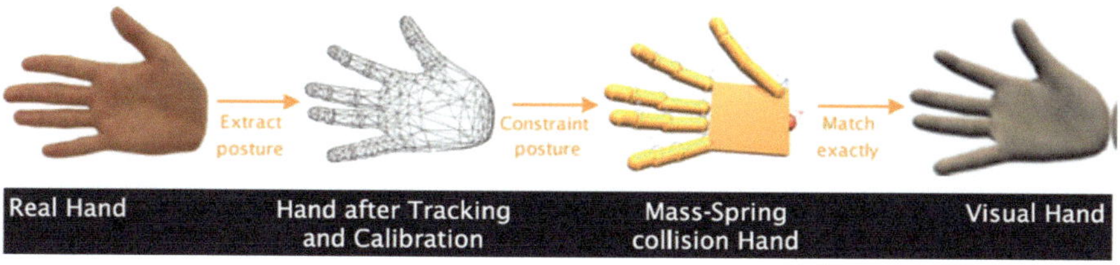

Fig. 9 The three hand models

- The Visual Hand, which is the only visible hand, using the visual rendering engine [12]. We couple each proxy hand with the tracked hand using virtual linear and angular springs. During the dynamic simulation, the spring-hand tends to follow the tracked-hand. The rendered hand visible for the user corresponds to the spring-hand configuration.

The "proxy" method we used has been first proposed for the Phantom (single-point interaction device), but, was extended to the full hand by Borst and Indugula [13] using the CyberGrasp™, which is a force feedback device also integrated in the Haptic Workstation™. The method solves the problem of interpenetration between the visual hands and the environment because the spring-hands adapt their pose on the surfaces of the objects. Spring-Hands have basically two constraints:

- Soft constraint: used to match the configuration of the tracked hands as much as possible by applying specific force and torques on the linear and angular springs.
- Hard constraint: used to avoid penetration within virtual objects by activating the collision detection between the phalanxes/palm rigid bodies and the objects of the Virtual Environment.

3.2 Wearable System for Mobility Improvement of Visually Impaired People

The system we have designed consists in sensing the surrounding environment via sonar sensors and sending vibro-tactile feedback to the user of the position of the closest obstacles in range [14]. The idea is to extend the senses of the user through a cyborgian interface. This means that the user should use it, after a training period, without any conscious effort, as an extension of its own body functions. Since there's reluctance from the visually impaired community toward new technologies, we design our system as a complement of the traditional white cane. It will focus on detecting obstacle at shoulder high and on letting the user perfectly hand free.

Some new commercial devices have appeared on the market, like the UltraCane [15] detects obstacles because it is equipped with a dual-range, narrowbeam ultrasound system that provides a 100% hazard protection envelope in front of and, uniquely, forward of the head and chest of the user. Two ultrasound transducers provide range data on the closest potential hazards, such as street furniture, plants, people, road signs, and over-hanging branches. This makes the UltraCane equally usable on the street as in interior spaces.

As shown in Fig. 10, a system architecture has been developed where its components are attached to a jacket. This wear cloth provides a natural way of carrying that facilitate its use. The components we integrate are: 4 sonar sensors (Fig. 11), a microcontroller, 8 vibrators, and a pc for the calibration needs. Other components are shown in Fig. 12.

3.3 *Enhancing Pilot Performance with a SymBodic System*

Increased fatigue of pilots during long flights can place both humans and machine at high risk. In this section, we describe our research on a SymBodic (SYMbiotic BODies) system designed to minimize pilot fatigue in a simulated 48 h mission. The system detected the pilot's sleep breaks and used this information to plan future

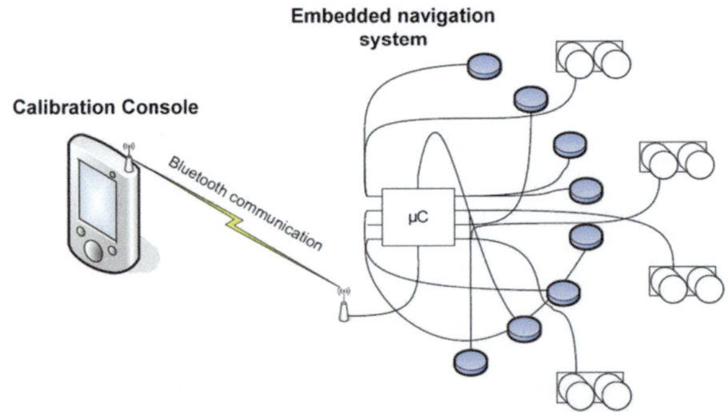

Fig. 10 Schema of the system architecture

Fig. 11 Sonar sensor

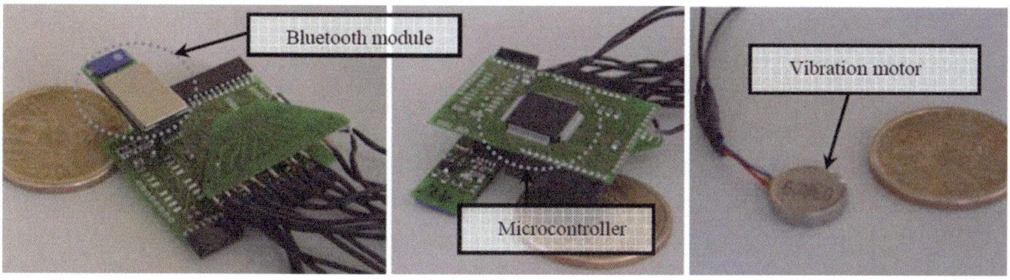

Fig. 12 Embedded system

sleep breaks. When fatigue could not be prevented, the SymBodic system assisted the pilot by providing relevant flight information through a vibro-tactile vest.

The SymBodic system [16] is composed of four major elements (Fig. 13): (A) A plane state recording system; (B) A pilot physiological recording system; (C) A central processing unit (CPU); and (D) A haptic feedback with a vibro-tactile vest.

The entire system is designed to be nonintrusive, light-weight, and with small energy consumption. The SymBodic system has been developed in the framework of the Solar Impulse development. Solar Impulse was an ambitious project to fly around the world with a solar powered airplane [17]. The project aimed to prove the concept of flying over long distances with renewable energies only.

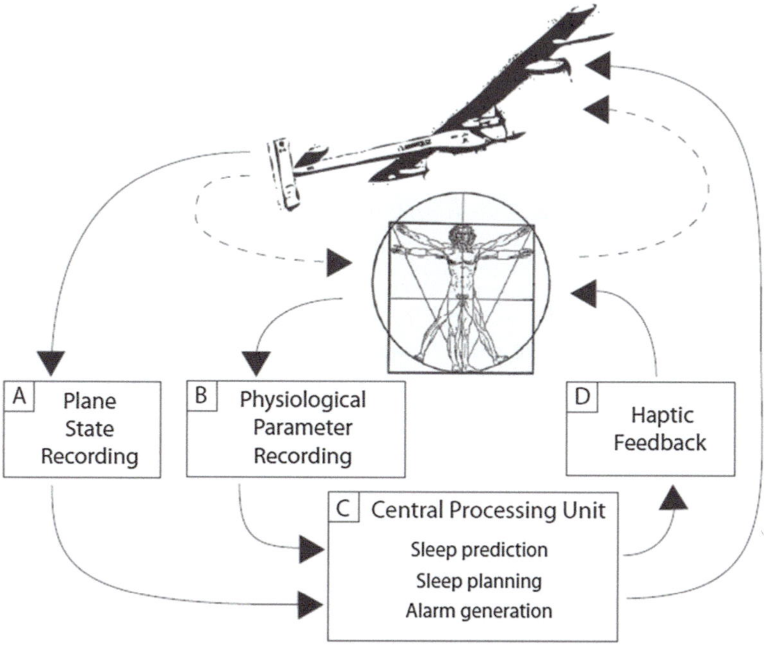

Fig. 13 The SymBodic system

(1) Plane state recording: Flight dynamics (roll, pitch and yaw) were recorded with the plane's internal sensors and were sent to the CPU. To test the prototype, we collected the flight dynamics from a realistic flight simulator called X-Plane (Laminar Research, SC).
(2) Physiological data recording: The physiological signals ECG and RSP were measured using a commercial available and certified recording system called Equivital (Hidalgo Ltd., UK). This wearable system is composed of a washable belt equipped with 3 dry textile electrodes and a piezo-resistive strain-gauge for the measurement of respiratory effort, and an electronics module for signal acquisition. The signals were sent over a serial Bluetooth wireless link to the CPU.
(3) CPU: A PC acquired the cardio-respiratory data and the flight dynamics. The cardio-respiratory data were used for the sleep/wake classification. The classification algorithm was adapted from [18]. Spectral features were extracted with a short time Fast Fourier Transformation from each consecutive 20 s long segment of the raw ECG and RSP signals. A single layer, feed-forward Artificial Neural Network classified the segments into wake and sleep. If the system detected that the pilot was sleeping longer than the 20 min required for the polyphasic sleep, an auditory wake up alarm was generated.
(4) Haptic feedback: Haptic feedback was provided with a vibrating vest. The vibrating vest was composed of 32 vibrating motors distributed around the torso of the pilot. The actuators were positioned to establish a morphological correspondence between the body of the pilot and of the machine [19]. The maximal intensity of each motor was calibrated with an external application to compensate for the perceptual difference due to variation of skin sensitivity at different body locations.

3.4 Actuators to Feel the Wind: Effective Versus Embedded

Righetti et al. [4] developed a tele-surveillance application using a remote controlled blimp (see Figs. 14 and 15). In early testing they encountered issues in piloting the remote blimp around various building. The wind at the passing of building's edges was changing drastically and induced tremendous deviation in the aircraft course. In order to collect information about this effect to the distant pilot, a set of relative pressure sensor were scattered around the blimp. Once the wind information is measured the system sends the data through wireless communication to an embedded system on the ground which renders this information around the waist via vibrotactile cues. The whole application shows improvement of the piloting condition by adding metaphorical extra-sensorial information about the local perturbation around the aircraft.

The main objective was to improve the immersion of the pilot, but the vibrotactile feedback tends to induce a break in presence. A better immersion was

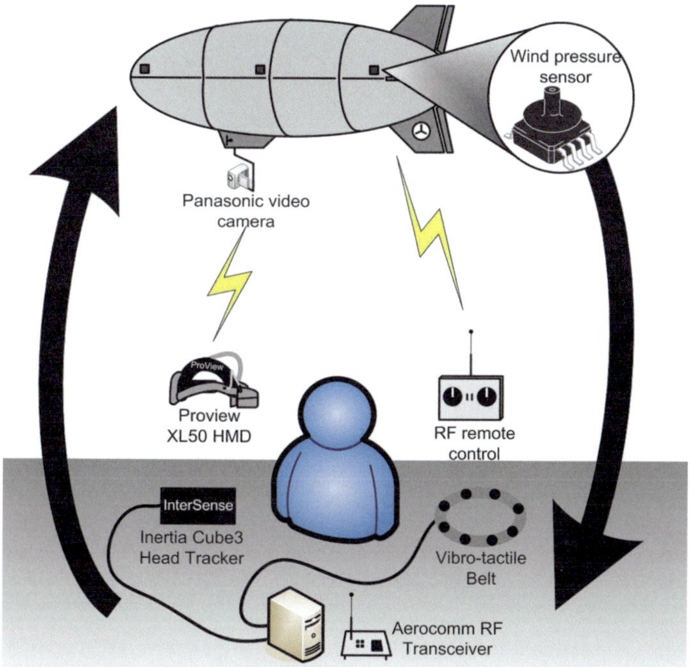

Fig. 14 The EPFL VRLab remote blimp system

Fig. 15 The EPFL VRlab blimp

obtained by passing the wind information to the user generating real wind around his head. He is able to feel the wind as if he was onboard the aircraft. The goal was to fill the gap between the needs of wind representation for tele-operation of UAV and the lack of available hardware, by presenting the development of a device which renders multidirectional wind on the user head (Fig. 16). The main idea was to fix a set of fans directly around the user head (see Fig. 17). In order to prevent limitation of the visual field the graphical feedback is given directly in front of the user eye using a Head-Mounted Display (HMD). Using a sufficient number of fans

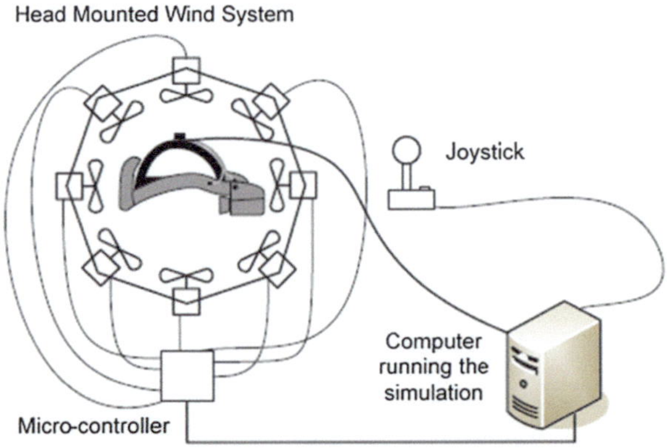

Fig. 16 The head-mounted wind system

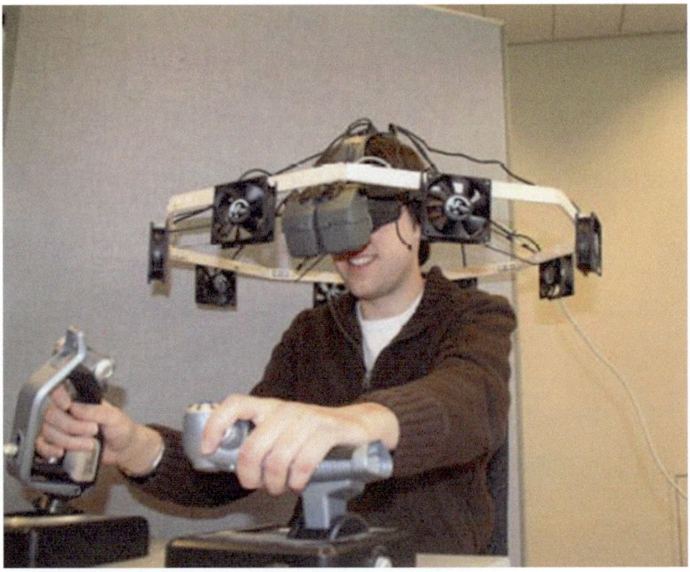

Fig. 17 Head-mounted wind

we can reproduce wind feedback in every possible direction. The wind force is rendered by tuning the fan speed.

In order to render omni directional wind we fix the number of fans to 8, equally distant from each other in every direction of the user head, as shown in Fig. 1.

4 Jackets

Since the early 2000s, the smart jacket has been lingering onto the market. Even so, it has yet to go mainstream. With new developments in enabling tech, the smart jacket trend is not going to disappear. More companies are providing garments with embedded functionalities like the "Know Where Jacket" with integrated GPS monitoring, or Bluetooth-enabled jacket from Burton and Motorola. O'Neill has launched in collaboration with MyWay, the NavJacket, with an interesting friend finder feature and a Bluetooth connection for displaying maps on the user's mobile phone. A designer from Brazil has also created some GPSenabled lingerie [20]. With this approach, if users want to change functionalities, they need to change their clothes. Alternately, MIThril, a complete modular wearable platform [21], has been developed to facilitate the development of distributed real-time multimodal and context-aware applications. Used in conjunction with the WearARM [22], a wearable computing core, researchers are trying to build a high-performance wearable system that is truly wearable and unobtrusive. On a more DIY approach, Leah Buechley has developed the LilyPad [23], based on the Italian Arduino platform. It can be sewn to the fabric and connected with various wearable sensors and actuators. Studio5050 [24] is providing a set of wearable building blocks to help rapidly develop wearable applications. Their modules communicate over a single communication line. Omius Tech [25] has studied the phenomenon of plants that can stay warm or cold without clothing on. The goal is to allow for the clothing to recognize when the body is hot or cold and adjust the air flow. Baubax has created a traveling jacket [26] that has specialized pockets for storing all of you most used electronics. Sinapsi [27] is a heated line jacket that uses Bluetooth to connect any iOS and Android devices. Once connected, three heating panel placed on the front and back along with multi-level heat adjusting battery keep you warm for up to 10 h. With the mobile app, we can schedule appointments on the jacket about where and when you want to go and be alerted when the weather conditions are changing. Another jacket developed by Thermal Tech [28] captures the power of the sun with the patented solar-absorbing technology that converts UV light (sun) into warmth to heat people up to 20° in just 2 min. ThermalTech is designed to keep people warmer longer, thus allowing them to do the things they love the most in the outdoors. Vodafone Netherlands [29] developed a smart jacket for cyclists; they can map out their route on a map app and then tuck their phone into a special pocket, which uploads the route and transmits it to a built-in Raspberry Pi device hooked up to 300 built-in LEDs. Lights on the jacket sleeves direct the traveler along the best route to take while more LEDs on the back of the jacket alert drivers about where they are heading. Emel and Aris propose a Smart Coat [30] with a slim and light-weight revolutionary heating system. The heat spreads through an inert specially formulated polymer.

We have introduced a novel modular approach towards wearable computing using I2C, a simple and proven two-wire communication protocol. For this, we

Fig. 18 Connector with magnets orientation and jacket

have designed and produced a set of small wearable modules in order to demonstrate in various occasions (EU project reviews, social events, workshops) the possibilities offered by our approach. Those modules consist of a master module, a bluetooth module, a WLAN module, a storage module, vibrotactile modules, accelerometer modules, a text-to-speech module and a LED module. In order connect the modules as easily and sturdily as possible, we have experimented with several techniques and have ended up using a unique combination of rivets and magnets for a usable and steady, plug-and-play system, as shown in Fig. 18. This modular approach has been used in the context of the INTERMEDIA EU project [31] to convey the vision of the user as multimedia central.

- As, the name suggests, the ***master module*** is the most important one as all slave modules will depend on it.
- An ***accelerometer module*** can be used for various purposes. For example, a user can place it on his/her wrist to specify input commands via hand gestures. Moreover, many accelerometer modules could be placed around the body and used to detect the posture and the movements of the user in order to derive his/her activity. The advantage of such inertial motion capture system over traditional visual or magnetic systems is that the user is not confined in a small area but is truly mobile. Such system has been successfully developed by Xsens5.
- In addition to the shared components cited previously, the ***vibrotactile module*** houses a small motor with an unbalanced mass on its driveshaft. The same vibrators can be found in mobile phones for example. Not only vibrators can inform us of an event (such as a phone call), but they also provide us with some additional information such as navigational cues for visually impaired people

[14]. Moreover, such vibrators can be used to enhance the user's perception as shown in [32]. For application for pilots, the position and the strength of the vibration will inform the user about the eventual deviation compared to the optimal trajectory and thus about the kind of correction that has to be applied.

- The hardware design of the ***LED module*** is exactly the same as the vibrotactile module except for a bright green LED that replaces the vibrator. Various commands can be sent to the module such as "light up at a specific brightness" or "blink with a specific interval". This LED module may be more intended for design and fashion, but could also be used for safety issues. A simple example of application that combines two accelerometer modules and two LED modules would be the case of LEDs that light up whenever a cyclist raises his arm when turning in one direction.
- A 3DOF ***Orientation Tracker*** provides drift-free 3D orientation as well as kinematic data: 3D acceleration, 3D rate of turn and 3D earth-magnetic field. This sensor will help us determine the deviation error according acceleration values calculated for the optimal trajectory in simulation.
- The ***text-to-speech module*** will announce the name of the manoeuvre to practice and when to start it. Moreover, it will inform the user about his error rate at the end of the manoeuvre. The module is connected with the onboard headphone system and thus improves the interaction with the user by giving auditory feedbacks. Without such feedback, Pirhonen et al. [33] have demonstrated that users perform interaction gestures worse than when audio feedback is provided.
- The ***storage module*** houses a microSD card reader in order to store data such as user preferences, GPS coordinates, addresses of nearby Bluetooth devices, interaction log, etc.
- The ***real-time clock module*** simply counts seconds, minutes, hours, date of the month, month, day of the week, and year with leap-year compensation.
- The ***bluetooth module*** has been designed to enable wireless communications between the wearable modules and a Bluetooth-enabled device such as a computer or a mobile phone. Our wearable Bluetooth module can be used interchangeably or in addition with the WLAN module as the communication type is transparent to the master module.
- ***Serial-to-WLAN IC modules*** offer the possibility to connect any embedded device to the internet, thus offering virtually unlimited of possibilities in terms of functionalities.
- The ***power module*** consists of a battery and a 5 V voltage regulator. We have attached a standard power connector in order to fit the battery inside the front pocket of the jacket. The capacity of the battery is enough to power the jacket for approximately a day depending on the number and kind of modules that are connected.

Figure 19 shows a few slave modules.

Fig. 19 Slave modules (from *left* to *right* storage module, vibrotactile module, accelerometer module

5 Conclusion

This paper has shown the importance of sensors and actuators in Human–Computer Interaction and especially in Virtual Reality and immersive games. We have emphasized the role of haptic and tactile feedbacks with four different case studies: haptic feedback using a haptic workstation, a wearable system for mobility improvement of visually impaired people, a system for enhancing pilot performance, and actuators to feel the wind. We have described a prototype of smart jacket with modular sensors and actuators.

Acknowledgements This research is supported by the Being Together Centre, a collaboration between Nanyang Technological University (NTU) Singapore and University of North Carolina (UNC) at Chapel Hill. The Being Together Centre is supported by the National Research Foundation, Prime Minister's Office, Singapore under its International Research Centres in Singapore Funding Initiative.

References

1. I.E. Sutherland, The ultimate display. Proc. IFIPS Congr. **2**, 506–508 (1965)
2. http://www.businessinsider.sg/facebook-oculus-plans-teleporter-by-2025-201511/?r=US&IR=T#CEKAX7JKym0e0FYw.97
3. H.G.M. Pakala, I. Khan, K. Raju, Sensors and actuators integration in embedded systems. ACEEE Int. J. Netw. Secur. **02**(02) (2011)
4. http://phys.org/news/2015-11-glove-cold-virtual-reality-apps.html#jCp
5. T. Molet, A. Aubel, T. Çapin, S. Carion, E. Lee, N. Magnenat Thalmann, H. Noser, I. Pandzic, G. Sannier, D. Thalmann, Anyone for tennis?, Presence, vol. 8, no. 2 (MIT, 1999), pp. 140–156
6. A. Aristidou, J. Lasenby, Motion capture with constrained inverse kinematics for real-time hand tracking, in *Proceedings of the 4th ISCCSP* (2010), pp. 1–5
7. Cyberglove 2, http://www.cyberglovesystems.com/

8. H. Liang, J. Yuan, D. Thalmann, Resolving ambiguous hand pose predictions by exploiting part correlations. IEEE Trans. Circ. Syst. Video Technol. **25**(7), 1125–1139 (2015)
9. L. Ge, H. Liang, J. Yuan, D. Thalmann, Robust 3D hand pose estimation in single depth images: from single-view CNN to multi-view CNNs, in *Proceedings of IEEE Conference on Computer Vision and Pattern Recognition (CVPR'16)* (2016)
10. R. Ott, F. Vexo, D. Thalmann, Two-handed haptic manipulation for CAD and VR applications. Comput. Aided Des. **7**(1), 125–138 (2010)
11. C. B. Zilles, J. K. Salisbury, A constraint-based god-object method for haptic display, in *Proceedings of the IEEE/RSJ International Conference on Intelligent Robots and Systems*, vol. 3 (1995), pp. 146–151
12. A. Peternier, D. Thalmann, F. Vexo, Mental vision: a computer graphics teaching platform, in *Proceedings of the 2006 Edutainment Conference* (2006), pp. 223–232
13. C.W. Borst, A.P. Indugula, Realistic virtual grasping, in *Proceedings of the 2005 IEEE Conference on Virtual Reality (VR'05)* (IEEE Computer Society, Los Alamitos, 2005), pp. 91–98
14. S. Cardin, F. Vexo, D. Thalmann, Wearable obstacle detection system for visually impaired people, in *VR Workshop on Haptic and Tactile Perception of Deformable Objects, Hannover, Germany, December 2005*
15. https://www.ultracane.com/
16. W. Karlen, S. Cardin, D. Thalmann, D. Floreano, Enhancing pilot performance with a symbodic system. in *Proceedings of the Engineering in Medicine and Biology Society (EMBC)* (IEEE, 2010), pp. 6599–6602
17. Solar Impulse (2010), http://www.solarimpulse.com
18. W. Karlen, C. Mattiussi, D. Floreano, Sleep and wake classification with ECG and respiratory effort signals. IEEE Trans. Biomed. Circ. Syst. **3**(2), 71–78 (2009)
19. S. Cardin, F. Vexo, D. Thalmann, Vibro-tactile interface for enhancing piloting abilities during long term flight. J. Robot. Mechatron. **18**(4), 381–391 (2006)
20. GPS Lingerie, http://bit.ly/Xhog4
21. R. DeVaul, M. Sung, J. Gips, A. Pentland, MIThril 2003: applications and architecture, in *Proceedings of the Seventh IEEE International Symposium on Wearable Computers* (2003), pp. 4–11
22. U. Anliker, P. Lukowicz, G. Troester, S.J. Schwartz, R.W. DeVaul, The WearARM: modular, high performance, low power computing platform designed for Integration into everyday clothing, in *Proceedings of the 5th IEEE International Symposium on Wearable Computers* (2001)
23. L. Buechley, M. Eisenberg, J. Catchen, A. Crockett, The LilyPad Arduino: using computational textiles to investigate engagement, aesthetics, and diversity in computer science education, in *Proceedings of the SIGCHI Conference on Human Factors in Computing Systems (CHI)* (2008), pp. 423–432
24. Studio5050, http://5050ltd.com/modules
25. http://www.omiustech.com/
26. https://www.kickstarter.com/projects/597538543/the-worlds-best-travel-jacket-with-15-features-bau
27. http://www.sinapsitecnologie.com/kickstarter/
28. https://www.indiegogo.com/projects/thermaltech-the-first-solar-powered-smart-jacket-solar#/
29. http://creativity-online.com/work/vodafone-smart-jacket/46067
30. http://emelandaris.com/
31. http://cordis.europa.eu/project/rcn/79765_en.html
32. X. Righetti, S. Cardin, D. Thalmann, F. Vexo, Immersive flight for surveillance applications, in *Proceedings of the IEEE Symposium on 3D User Interfaces* (2007) pp. 139–142
33. A. Pirhonen, S. Brewster, C. Holguin, Gestural and audio metaphors as a means of control for mobile devices, in *Proceedings of the SIGCHI conference on Human Factors in Computing Systems: Changing Our World, Changing Ourselves* (ACM, 2002), p. 298

Chapter 5
Frequency Conversion Layers for Si Solar Cell Efficiency Improvement

Fabrice Gourbilleau, Lucile Dumont, Julien Cardin, Anaïs Gouesmel, Ing-Song Yu, Christophe Labbé and Hocine Merabet

Abstract SiNx and SiNx: Tb^{3+} thin layers deposited by reactive magnetron co-sputtering have been studied with the aim of optimizing the light management in Si solar cells. Those Si-based layers are developed to be compatible with the Si-PV technology. An efficient energy transfer between matrix and terbium ions has been demonstrated and optimized. The layer composition and microstructure as well as its optical properties have been analyzed to favor the required optical properties for achieving a good solar cell, i.e., good anti-reflective properties and high luminescence emission intensity. Finally a Tb-doped SiNx thin film has been deposited on the top of a Si solar cells and solar cell characteristics studied.

Keywords Si solar cell · Frequency conversion · Down conversion · Down shifting · Silicon nitride · Alumina

Managing the energy harvesting is one of the key issues for this century. Many efforts have been devoted these last years to develop innovative solutions for (i) decreasing the CO_2 production (ii) limiting the energy consumption, (iii) increasing the renewable energy use. Among these solutions proposed, such as energy storage, energy conversion, fuel cells, cooling systems, the development of luminescent materials with high efficiency keeping a low cost process is one of the challenges to face. Luminescent materials are well known to be widely used in light

F. Gourbilleau (✉) · L. Dumont · J. Cardin · A. Gouesmel · C. Labbé
CIMAP, Normandie Université, ENSICAEN, UNICAEN, CEA, CNRS,
14000 Caen, France
e-mail: fabrice.gourbilleau@ensicaen.fr
URL: http://cimap.ensicaen.fr/

I.-S. Yu
Department of Materials Science and Engineering, National Dong Hwa University,
Hualien, Taiwan, China

H. Merabet
Department of Mathematics, Statistics, and Physics,
College of Arts and Sciences, Qatar University, Doha, Qatar

S.R.S. Prabaharan et al. (eds.), *Frontiers in Electronic Technologies*,
Lecture Notes in Electrical Engineering 433, DOI 10.1007/978-981-10-4235-5_5

emitting diodes, sensors, scintillators, as well as in photovoltaic applications. In most cases, the photon emission is obtained taking benefit of rare earth ion excitation mechanism that offers a wide range of emission wavelengths ranging from the UV to the infrared.

For the case of Silicon solar cell, increasing its efficiency while keeping a low cost process is one of the goals of the Si-PV industry to continuously decrease the cost of the power generation. This is the industrial condition to remain a major player in the provision of power generation solutions in the forthcoming years. The different loss paths in a Si solar cell are well identified and many developments have been performed to solve them. Among them, the thermalization mechanism that is the consequence of the mismatch between the solar spectrum energy (UV region) and the solar cell band gap energy (1.1 eV of the Si solar cell) can be overcome. Such an objective can be achieved by using frequency conversion layers so-called Down Conversion (DC) [1–4] or Down Shifting (DS) [1, 5] layers. Such layers have been developed in order to convert one UV incident photon into two IR ones (DC case) or one UV incident photon into one visible one (DS case) that can be absorbed by the Si cell. To reach this objective many systems using a couple of trivalent ions such as Pr^{3+}-Yb^{3+}, Tb^{3+}-Yb^{3+}, Ce^{3+}-Yb^{3+} for DC process or one trivalent ions such Pr^{3+}, Tb^{3+} for the DS conversion have been studied. Unfortunately, the major drawbacks of these layers are the use of a non Si-compatible process due to the nature of the host matrix as well as the low absorption cross section of the rare earth ions that limits their excitability in the solar spectrum range [5–9]. To overcome this problem, a Si-PV compatible host matrix containing sensitizers has been developed. The presence of sensitizers allows to efficiently excite rare earth ions. Moreover, to keep a low cost process for a future development in the SI-PV industry, the host matrix developed should have good anti-reflective properties.

This paper details the development a Si-PV compatible host matrix that efficiently absorbs the solar spectrum and allows an efficient excitation of rare earth ions. Moreover, such a host matrix should favor the incorporation of a high content of rare earth ions without the detrimental clustering effect [10, 11]. The purpose of this paper is to describe the fabrication and study of Tb:Yb and Tb-doped Si-based thin films or Al-based thin films deposited by reactive co-sputtering or ALD techniques, respectively.

The layers have been deposited on p-type 250 μm-thick [001] 2'' silicon (Si) substrates by either reactive magnetron co-sputtering in a nitrogen-rich plasma or Atomic Layer Deposition using an oxygen plasma. For the former, the matrix composition (Si/N ratio) was tuned by varying the Ar/N_2 ratio of the gas flux injected or the plasma pressure while for the later, the Al_2O_3-based layer characteristics have been monitored through the Oxygen Plasma time and the substrate temperature. For the sputtering approach, the rare earth ions incorporation was controlled by the RF power density applied on the Tb^{3+} target (RFP_{Tb}). RFP_{Tb} was varied between 0.3 and 1.8 W/cm^2. After optimization, Yb ions have been incorporated to produce the DC layers. For all the undoped- and Tb^{3+}- and Tb^{3+}-Yb^{3+} doped -SiN_x deposited layers, the target power density applied on the silicon target was fixed at 4.5 W/cm^2 in agreement with previous studies [12]. The deposition temperature was set at 200 °C while the deposition time was adjusted to obtain 90 nm-thick films whatever the deposition conditions. After deposition, the samples were annealed during 1 h by classical thermal annealing (CTA) at 850 °C. Concerning the ALD deposition techniques, only DS layers have been fabricated using precursors containing Tb. The fabricated layers have been annealed after deposition at lower temperature (<650 °C) using either classical thermal annealing or rapid thermal annealing approaches. The composition of the fabricated layers has been investigated by means of Fourier Transform Infrared measurements (FTIR). They were performed at room temperature thanks to a Thermo Nicolet Nexus 750 II spectrometer working in the 4000–400 cm^{-1} range, with a resolution of 5 cm^{-1}. The emission and excitation properties, photoluminescence (PL) and photoluminescence in excitation (PLE) experiments were performed at room temperature on the films. PL experiments have been carried out using a Lot-Oriel 1 kW-Xenon lamp connected to an OMNI300 monochromator. The PL spectra were recorded with a Hamamatsu (R5108) photomultiplier tube after the dispersion of the PL signal by a MSH 300 OMNI monochromator. The detection system was locked in with a SR830 amplifier referenced at the excitation light beam chopped frequency. The same system has been used for the PLE experiment. PL and PLE spectra obtained corrected by the set-up and by the lamp emission intensity in the PLE case.

First Tb-doped systems have been optimized to get the maximum photoluminescence emission under UV range excitation for DS process or prior to incorporating the Yb^{3+} ions. An example of the achieved Tb emission in a Tb-doped SiN_x host matrix fabricated by reactive co-sputtering is shown on Fig. 1a. The same feature is achieved for the Tb-doped Al_2O_3 layers obtained by ALD. Four emission peaks can be observed corresponding to the transitions from the $^5D_4$ to the ^7F_i levels as presented on Fig. 1b. Note that the emission spectrum has been achieved under a low photon flux (1.9 × 10^{15} photons/cm^2) excitation at 300 nm for the Tb^{3+} ions which has a low absorption cross section below 2 × 10^{-21} cm^2 as reported in [13]. Such a result evidences an efficient indirect excitation of Tb^{3+} ions which may witness a sensitization of the Tb^{3+} ions by defect states of the matrix as already evidenced in Tb^{3+}-doped SiO_xN_y and Tb-Si_xO_y matrices [10].

With the incorporation of Yb^{3+} ions to the system, we observed a decrease of the Tb emission while the one corresponding to the $^2F_{5/2}$-$^2F_{7/2}$ transition (980 nm) of the Yb^{3+} ions rises. This is the signature of an efficient energy transfer from the Tb^{3+} ions to the Yb^{3+} ones. For this system, a quantum efficiency as high as 200% has been achieved in a composite or a multilayer structure. To improve the coupling rate between the rare earth ions, fabrication parameters have been optimized. The goal is to get the maximum of emitted photon at 980 nm. After optimization, these frequency conversion layers have been deposited on top of industrial crystalline Si solar cells. For this purpose, their deposition steps have been integrated in the industrial cells fabrication process.

Reflectance measurements have been carried out on the different frequency conversion layers before depositing the metallic contact for the achievement of the Solar Cell. A typical reflectance measurement result is displayed on Fig. 2. The

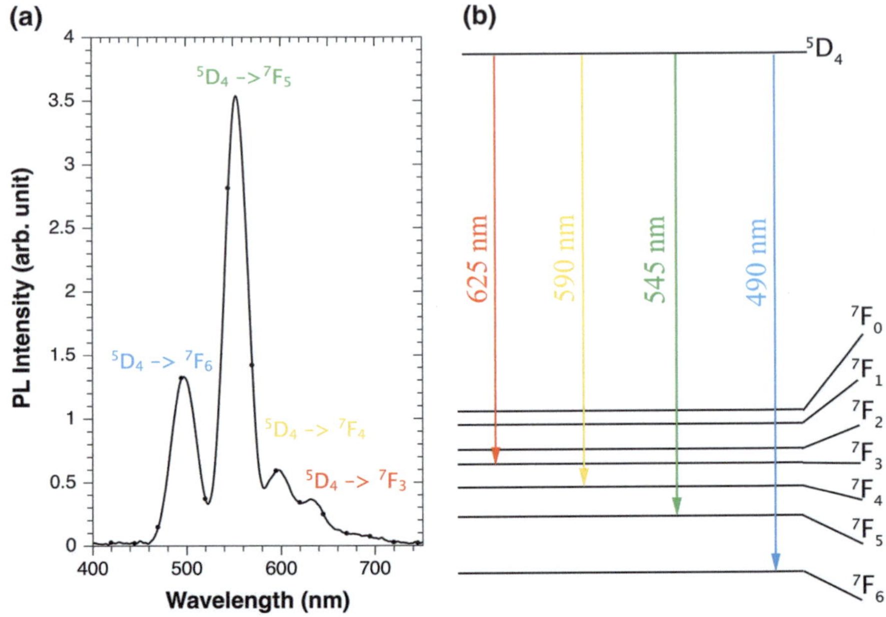

Fig. 1 **a** PL spectrum of a Tb-doped SiN_x under 300 nm excitation wavelength. **b** Energy level scheme of the Tb^{3+} ions

reflectance values for a nitride layer used as a reference and the Tb-doped layers are below the value of 5% in wide domain of wavelength ranging from 350 to 750 nm and a bump is observed after 1020 nm. The reflection efficiency in the 300–1200 nm range is of 6.04% for the device coated with the SiN_x layer and reaches 6.20% by adding the rare earth ion. Such value demonstrates the high anti-reflective properties of the SiN_x host matrix whatever it is doped or not.

External and Internal Quantum Efficiency measurements have been carried out on Si Solar Cell for these Down Shifting- and the Down Conversion- layers. A typical Internal Quantum Efficiency (IQE) of the solar cell on the top of which the SiN_x-based layers described in Fig. 2 have been deposited is displayed on Fig. 3.

The *IQE* spectra reveals that the Tb-doped layer has a higher *IQE* than the undoped one from 300 to 900 nm with a particular increase of 15% in the *UV* range (300–400 nm). Above 900 nm one can note a slight decrease of the doped layer with respect to the undoped one. This result shows that the *IQE* is larger for *UV* photons thanks to the Tb-doped layer.

Thus this latter acts as a down-shifting layer that may improve the solar cell efficiency. Such down-shifting layer can be a promising solution for improving the Si Solar Cell considering that the process developed is fully compatible with the industrial process. More details will be given on the electrical and optical

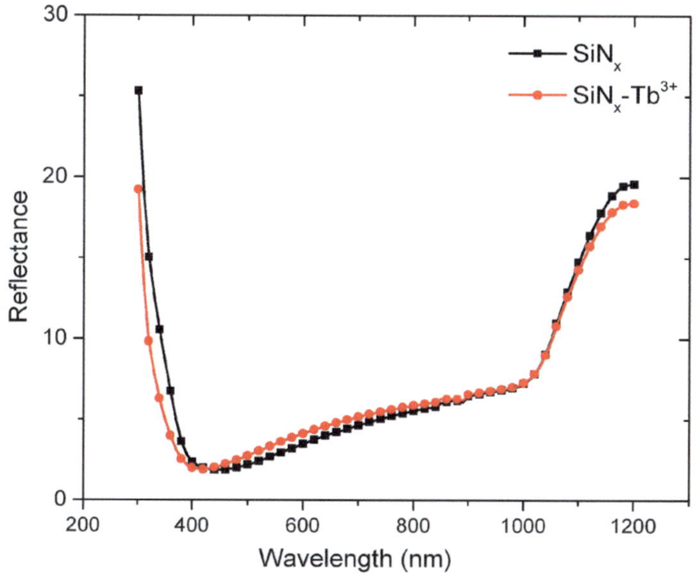

Fig. 2 Reflectance characteristics of the SiN_x (*black*) and SiN_x:Tb^{3+} (*red*) anti-reflective layers deposited on a textured Si surface

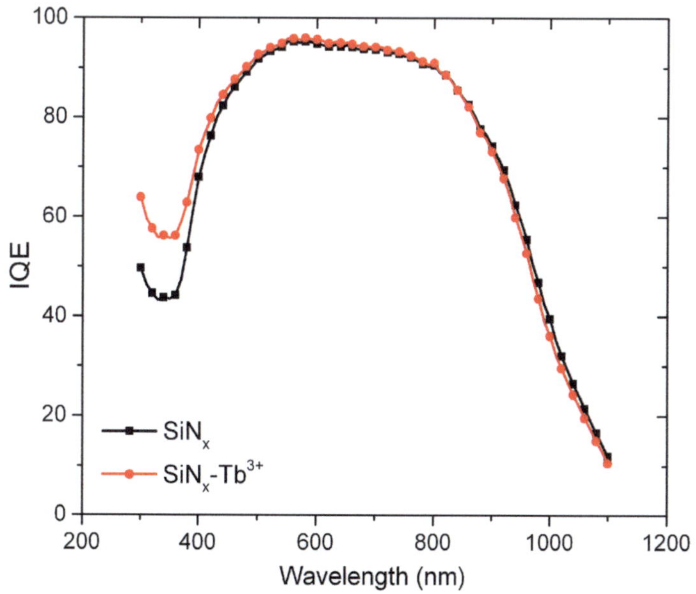

Fig. 3 IQE curves of a solar device with either the optimized SiN_x (*black*) or the optimized SiN_x: Tb^{3+} (*red*) layers on top

characteristics of these different frequency conversion layers and the increased efficiency achieved.

To conclude, in this paper, undoped, Tb^{3+}- and Tb^{3+}-Yb^{3+} doped SiN_x layers deposited by reactive magnetron co-sputtering have been studied in order to obtain a thin DS or DC layer that have good anti-reflective properties and favor the emission of NIR photon that can be absorbed by the solar cell. For the ALD deposited layer, only DS layer (Tb^{3+}-Al_2O_3) layers have investigated. Studies of the composition, the microstructure, and the optical properties have been carried out to get the maximum emission efficiency of the rare earth ions. The IQE of the optimized layers deposited by means of co-sputtering approach were compared. Thus it was found that the DS layer improve by 15% the IQE in the UV part of the solar spectrum and by 25% the global efficiency of the Si solar cell. Concerning the DC layer, a relative increase of 1.5% has been evidenced demonstrating the promising potential of these Si-PV compatible layer for enhancing the efficiency of Si Solar Cell.

Acknowledgements This work has been supported by the French Research National Agency through the project GENESE (N° ANR-13-BS09-0020-01), by the French Ministry of Research through the ORCHID PHC project n°33572XF and the Qatar National Research Fund though the project Grant 8-1467-1-268.

References

1. C. Strümpel, M. McCann, G. Beaucarne, V. Arkhipov, A. Slaoui, V. Švrček, C. del Cañizo, I. Tobias, Modifying the solar spectrum to enhance silicon solar cell efficiency—an overview of available materials. Solar Energy Mater. Solar Cells **91**, 238–249 (2007). doi:10.1016/j.solmat.2006.09.003
2. B.S. Richards, Enhancing the performance of silicon solar cells via the application of passive luminescence conversion layers. Solar Energy Mater. Solar Cells **90**, 2329–2337 (2006). doi:10.1016/j.solmat.2006.03.035
3. K.D. Oskam, R.T. Wegh, H. Donker, E.V.D. van Loef, A. Meijerink, Downconversion: a new route to visible quantum cutting. J. Alloys Comp. **300–301**, 421–425 (2000). doi:10.1016/S0925-8388(99)00755-0
4. B.S. Richards, Luminescent layers for enhanced silicon solar cell performance: down-conversion. Solar Energy Mater. Solar Cells **90**, 1189–1207 (2006). doi:10.1016/j.solmat.2005.07.001
5. E. Klampaftis, D. Ross, K.R. McIntosh, B.S. Richards, Enhancing the performance of solar cells via luminescent down-shifting of the incident spectrum: a review. Solar Energy Mater. Solar Cells **93**, 1182–1194 (2009). doi:10.1016/j.solmat.2009.02.020
6. K.R. McIntosh, G. Lau, J.N. Cotsell, K. Hanton, D.L. Bätzner, F. Bettiol, B.S. Richards, Increase in external quantum efficiency of encapsulated silicon solar cells from a luminescent down-shifting layer. Prog. Photovolt. Res. Appl. **17**, 191–197 (2009). doi:10.1002/pip.867
7. D. Chen, Y. Wang, Y. Yu, P. Huang, F. Weng, Quantum cutting downconversion by cooperative energy transfer from Ce3+ to Yb3+ in borate glasses. J. Appl. Phys. **104**, 116105 (2008). doi:10.1063/1.3040005

8. Q. Duan, F. Qin, D. Wang, W. Xu, J. Cheng, Z. Zhang, W. Cao, Quantum cutting mechanism in Tb3+-Yb3+co-doped oxyfluoride glass. J. Appl. Phys. **110**, 113503 (2011). doi:10.1063/1.3662916
9. X. Liu, S. Ye, Y. Qiao, G. Dong, B. Zhu, D. Chen, G. Lakshminarayana, J. Qiu, Cooperative downconversion and near-infrared luminescence of Tb3+–Yb3+codoped lanthanum borogermanate glasses. Appl. Phys. B. **96**, 51–55 (2009). doi:10.1007/s00340-009-3478-z
10. Y.-T. An, C. Labbé, M. Morales, P. Marie, F. Gourbilleau, Fabrication and photoluminescence properties of Tb-doped nitrogen-rich silicon nitride films. Phys. Status Solidi C **9**, 2207–2210 (2012)
11. L. Dumont, J. Cardin, P. Benzo, M. Carrada, C. Labbé, A.L. Richard, D.C. Ingram, W.M. Jadwisienczak, F. Gourbilleau, SiNx:Tb3+-Yb3+, an efficient down-conversion layer compatible with a silicon solar cell process. Solar Energy Mater. Solar Cells **145**, 84–92 (2016). doi:10.1016/j.solmat.2015.09.031
12. O. Debieu, R.P. Nalini, J. Cardin, X. Portier, J. Perrière, and F. Gourbilleau, Structural and optical characterization of pure Si-rich nitride thin films. Nanoscale Res. Lett. **8**(1), 31 (2013).
13. D.K. Sardar, K.L. Nash, R.M. Yow, J.B. Gruber, U.V. Valiev, E.P. Kokanyan, Absorption intensities and emission cross sections of TB3+ (4f8) in TbAlO3. J. Appl. Phys. **100**, 083108 (2006). doi:10.1063/1.2358401

Chapter 6
Design and Fabrication of 320 × 256 Focal-Plane Array Using Strain-Coupled Quaternary Capped InAs/GaAs Quantum Dots Infrared Photo-Detectors for Thermal Imaging

H. Ghadi, H. Rawool, K.C. Goma Kumari and Subhananda Chakrabarti

Abstract We report the fabrication and characterization of a 320 × 256 infrared focal-plane imager fabricated using an strain-coupled quaternary capped InAs quantum dots heterostructure, which showed multiple photoluminescence peak and activation energy of 207.38 meV for dominant peak. Multiple ground state peaks in photoluminescence spectra indicates multimodal dot size distribution which was confirmed using cross-sectional transmission microscopy images. We discuss the fabrication and characterization of single-pixel detectors that can measure intersubband spectral responses with peak intensity at 6.9 μm and narrow spectral linewidth of 19%. The highest detectivity of 2.48×10^{10} cm $Hz^{1/2}$/W at 77 K was observed from proposed structure. Using the fabricated device, infrared images were captured at 50–100 K. Device optimization led to approximately 95% of the pixels in the imaging array being operational and a reasonably low noise equivalent temperature of approximately 0.080 °C at 100 K.

Keywords Quaternary · Quantum dots · Focal-plane arrays

1 Introduction

Infrared (IR) imaging systems that use quantum well infrared photodetector–based and quantum dot infrared photodetectors (QDIP)-based large arrays in long wavelength IR and very long wavelength IR regions are used worldwide for monitoring weather profiles, resources, deforestation, and distribution of minor atmospheric constituents [1, 2]. The optical and electrical characteristics of

H. Ghadi · H. Rawool · K.C. Goma Kumari · S. Chakrabarti (✉)
Department of Electrical Engineering, Indian Institute of Technology Bombay, Mumbai 400076, Maharashtra, India
e-mail: subhanandachakrabarti@gmail.com; subho@ee.iitb.ac.in

S.R.S. Prabaharan et al. (eds.), *Frontiers in Electronic Technologies*,
Lecture Notes in Electrical Engineering 433, DOI 10.1007/978-981-10-4235-5_6

self-assembled quantum dot (QD) nanostructures closely match those of bulk structures [3]. A QDIP is a device that exploits the properties of QDs to realize advantages unavailable when using other semiconductor heterostructures. Compared with QWIPs, QDIPs exhibit higher responsivity because of the longer lifetime of the photo-excited carriers and the lower dark current [4, 5]. QDIPs couple the incident photon radiation to transfer carriers in the conduction band from the ground state to the excited state. QDs exhibit complete discrete energy levels within the CB and valance bands because of their 3D quantum confinement. The transitions between the inter-subbands can be utilized for IR detection [3–5].

A focal-plane array (FPA) is an image-sensing device that comprises an array of light-sensing pixels (i.e., single detectors) at the focal plane of a lens [6]. FPA performance depends on four factors: growth and processing of the detector material, design of the readout circuit, hybrid-ization of large FPAs, and packaging. 320 × 256 FPA is a hybrid array i.e. a combination of a detector and a Si CMOS readout integrated circuit (ROIC) [6, 7]. Instead of wire-bonding, the photocurrent from each pixel is collected using the In bumps grown on each pixel and the ROIC. Two arrays are carefully aligned and pressed using a flip-chip bonder. The In bump acts as an electrical connection between the detector material and the readout, thus collecting photocurrent from each pixel. The space between two arrays is filled with epoxy to structurally support the detector material. Finally, the detector is thinned to a thickness of only 10 μm (back illuminated) to increase absorption (i.e., quantum) efficiency.

2 Experimental Details

Growing QDs that meet the desired spectral performance is the first step in realizing 320 × 256 focal plane array. The device heterostructure (Fig. 1) was grown on a semi-insulating GaAs (100) wafer, on which InAs QD layers were grown at a rate of 0.2 monolayer (ML)/s at 520 °C through solid-source molecular beam epitaxy. Next, 2.7-ML InAs QDs were grown in the active region, followed by capping with the 30-Å quaternary $In_{0.21}Al_{0.21}Ga_{0.58}As$ and GaAs layers at 590 °C. The capping thickness of the GaAs layer was fixed to 120 Å. Ten periods of the active layer (QD with combination capping) were capped with a 1000-Å GaAs layer, and this stack was repeated three times. The whole structure was enclosed in high doped N+ layer. Bottom N+ thickness was 7000 Å and top N+ thickness was 2000 Å, respectively.

Optical characterization was carried out using photoluminescence in which samples were excited using a 25-mW, 532-nm diode-pump solid-state laser. Single pixel detectors were fabricated using 2 stage lithography process to evaluate electrical, spectral and blackbody characteristics of grown heterostructure. Focal-plane array (FPA) development is the next critical step, followed by hybridization using readout integrated circuit (ROIC) and indium bumps. An optimized fabrication process was used for fabricating the arrays. Samples were

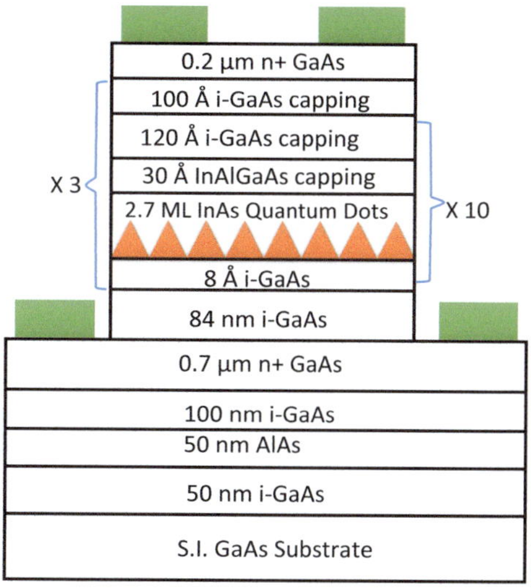

Fig. 1 Sample heterostructure grown using MBE

flip-chip bonded using bump-fabricated ROICs, which were thinned in the later stages, leaving behind only detector pixels, the bottom contact layer, and the AlAs layer. Wafer thinning was performed through lapping and polishing, wet etching to remove surface roughness, and dry etching to the epitaxially grown etch-stop layer. Detectors were thinned to reduce thermal mismatch between the ROIC and the strain-coupled detector array, pixel outages, and pixel-to-pixel optical crosstalk within the fabricated FPA. Further, a thinner substrate act as a waveguide that enhances the optical field, thus increasing quantum efficiency (QE).

Figure 2 depicts the fabrication process flow, including the fabrication and testing of a hybrid flip-chip array mounted on a commercially available ROIC, for a 320 × 256 pixel FPA.

3 Results and Discussion

Figure 3 shows the photoluminescence (PL) spectra recorded at 8 K with laser power of 100 μW and 25 mW. Multiple ground-state peaks were observed, affirming the multimodal QD size distribution. For smaller capping thicknesses, compressive strain arising from lattice mismatch propagates to a larger extent in the upper QD layers, resulting in larger QDs.

Multimodal dot size distribution was confirmed with cross-sectional transmission microscopy images (Fig. 4). Quantum dots were found to be vertically aligned and increase in dot size was observed compared to seed layer (bottom dot layer). The elastic strain propagating from the underlying QD layers along the growth direction thermodynamically favors [8] QD formation in the upper layers, resulting in site nucleation, narrow emission linewidth, increased intersubband absorption,

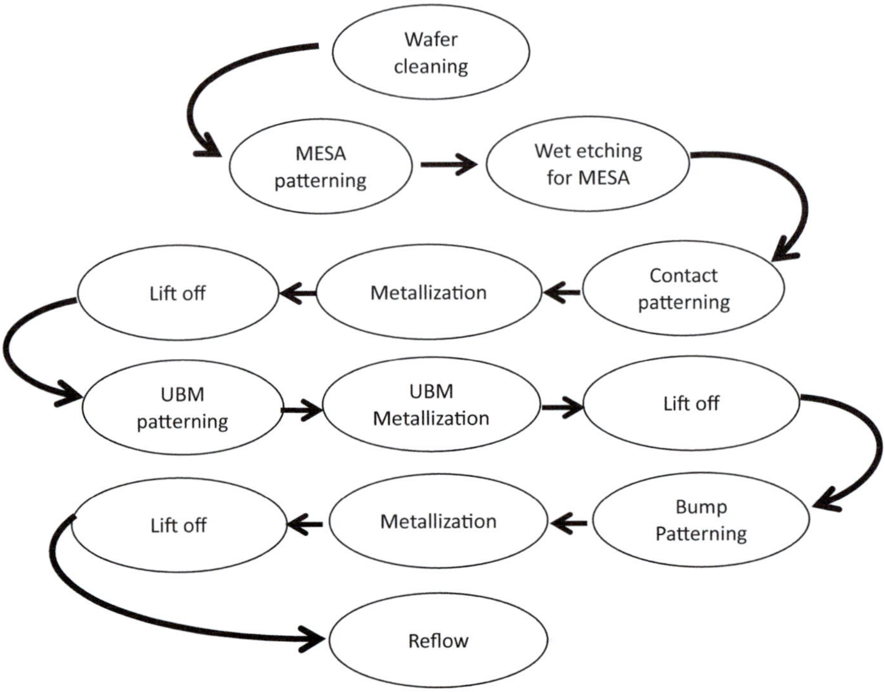

Fig. 2 Fabrication process flow for fabricating 320 × 256 FPA

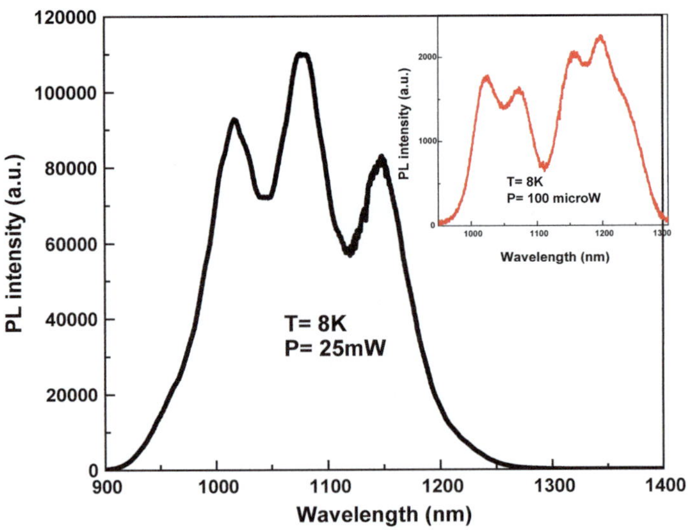

Fig. 3 Low temperature PL spectra recorded at 8 K and 25 mW laser power. *Inset* PL spectra recorded at same temperature and 100 mW power

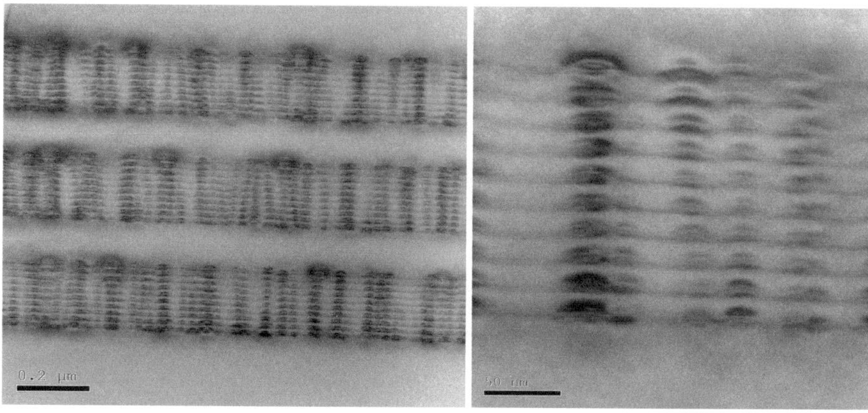

Fig. 4 XTEM images captured at 0.2 μm and 50 nm resolution

and potential improvement in device efficiency [9–11]. However, increased strain propagation from the underlying layers due to the decreased capping thickness in coupled QDs increases the possibility of defects and dislocations in the QD heterostructure. X-TEM images confirms the grown heterostructures were defects and dislocation free.

The activation energy (E_a) and optical properties of the heterostructure were examined through temperature-dependent PL analysis. The carrier escape mechanism was examined using diode equation (Eq. 1) [12].

$$I_D = I_{D0}\ \exp\ (E_a/KT) \tag{1}$$

where I_D is the PL intensity, and I_{DO} is the maximum PL intensity at absolute low temperature, i.e., at 8 K, K is the Boltzmann constant, and T is the temperature. E_a calculated using temperature-dependent PL measurements for proposed heterostructure was 207.38 meV. Blackbody and current voltage measurements were carried out to calculate detectivity and responsivity from fabricated single pixel detectors. Sample exhibited low dark current density with high peak detectivity (2.48 × 10^{10} cm $Hz^{1/2}$/W) measured at 77 K. The spectral response obtained at −1 V for fabricated device is shown in Fig. 5. The spectral width ($\Delta\lambda/\lambda$) calculated for the longest wavelength peak (where $\Delta\lambda$ is the full width at half maxima of the peak) was 19% [11, 12] The long wavelength peak is predicted to be due to a transition from the QD ground state to an excited state within the QD.

320 × 256 focal plane array was fabricated using proposed heterostructure and flip chip bonded to ROIC. The device that was flip-chip bonded with ROIC and wirebonded on LCC exhibited the optimal thermal imaging performance (95% yield). The wirebonded samples were loaded in a CamIRa system for thermal imaging. A low noise equivalent temperature difference of 0.080 °C was achieved for the fabricated FPA. Thermal images were observed up to 120 K; the optimal result is presented in Fig. 6.

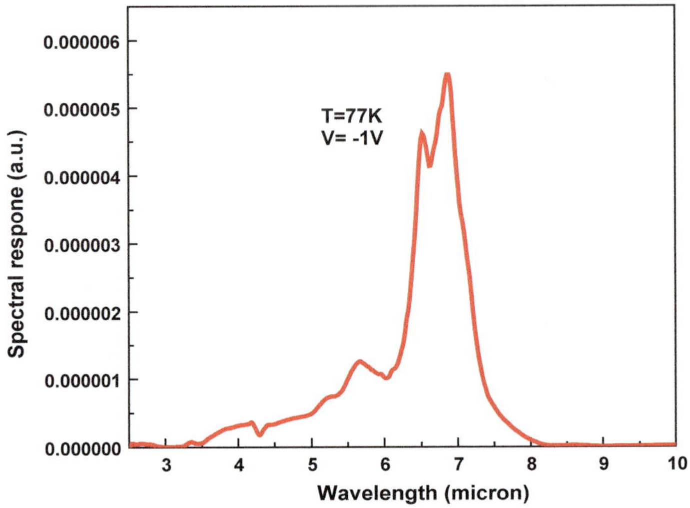

Fig. 5 Spectral response observed at 77 K with applied −1 V

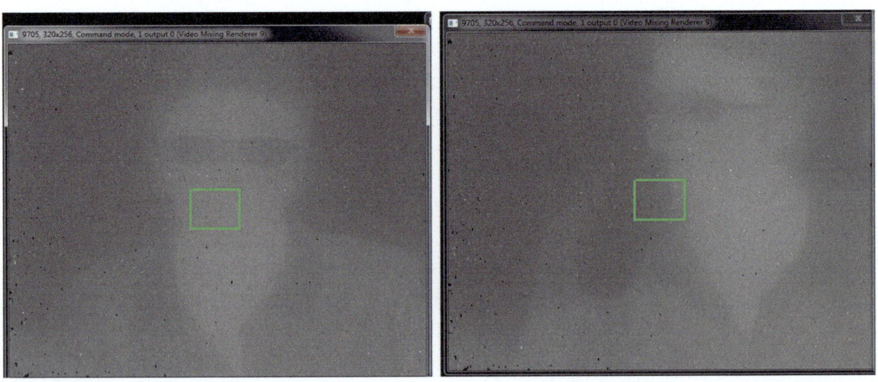

Fig. 6 Thermal image of a human observed at 100 K and an applied bias of 1.2 V

4 Conclusion

We report a successful development and fabrication of 320 × 256 focal plane imager developed using strain-coupled quaternary capped InAs/GaAs heterostructure. Low-temperature PL spectra showed multimodal dot size distribution. Vertically co-related dots with increase in dot sizes was confirmed using cross-sectional transmission microscopy images. Single-pixel detectors exhibited a dominant peak at 6.9 μm with 19% spectral line width. Fabricated detectors had low dark current and highest peak of detectivity of 2.48×10^{10} cm $Hz^{1/2}/W$ at 77 K. We have successfully fabricated a 320 × 256 nanotechnology based focal plane array capable of capturing images of human targets. An optimization process resulted in more than 95% of the pixels being operational with a reasonably low noise equivalent temperature of 0.08 °C at 100 K.

Acknowledgements The authors acknowledge the financial support provided by the National Center of Excellence in Technology for Internal Security (NCETIS), Department of Science and Technology (DST) Nanomission, Indian Space Research Organization (ISRO) and DST, India. We would also like to acknowledge IIT-Bombay's Nanofabrication Facility and Riber, France.

References

1. S. Tsao, H. Lim, W. Zhang, M. Razeghi, High operating temperature 320 × 256 middle-wavelength infrared focal plane array imaging based on an InAs/InGaAs/InAlAs/InP quantum dot infrared photodetector. Appl. Phys. Lett. **90**, 201109 (2007)
2. J. Jiang, K. Mi, S. Tsao, W. Zhang, H. Lim, T. O'Sullivan, T. Sills, M. Razeghi, G.J. Brown, M.Z. Tidrow, Demonstration of a 256 × 256 middle-wavelength infrared focal plane array based on InGaAs/InGaP quantum dot infrared photodetectors. Appl. Phys. Lett. **84**, 223 (2004)
3. P. Bhattacharya, S. Ghosh, A.D. Stiff-Roberts, Quantum dot opto-electronic device. Annu. Rev. Mater. Res. **34**, 1–40 (2004)
4. V. Ryzhii, The theory of quantum-dot infrared phototransistors. Semicond. Sci. Technol. **11**, 759–765 (1996)
5. S.D. Gunapala, S.V. Bandara, C.J. Hill, D.Z. Ting, J.K. Liu, S.B. Rafol, E.R. Blazejewski, J. M. Mumolo, S.A. Keo, S. Krishna, Y.-C. Chang, C.A. Schott, 640 × 512 pixels long-wavelength infrared (LWIR) quantum-dot infrared photodetector (QDIP) imaging focal plane array. IEEE J. Quant. Elec. **43**, 230–237 (2007)
6. T. Sprafke, J.W. Beletic, High performance infrared focal plane arrays for space application. Opt. Photonics News **9** (2008)
7. A.G. Stern, Design of high quantum efficiency and high resolution, Si/SiGe avalanche photodiode focal plane arrays using novel, back-illuminated, silicon-on-sapphire substrates. Science Technology and Medicine (2011)
8. B. Jogai, Three-dimensional strain field calculations in coupled InAs/GaAs quantum dots. J. Appl. Phys. **88**(9), 5050–5055 (2000). L.H. Li, M. Rossetti, G. Patriarche, A. Fiore, Growth of InAs bilayer quantum dots for long-wavelength laser emission on GaAs. J. Cryst. Growth **301**, 959–962 (2007)
9. L.H. Li, M. Rossetti, G. Patriarche, A. Fiore, Growth of InAs bilayer quantum dots for long wavelength laser emission on GaAs. J. Cryst. Growth **301**, 959–962 (2007)
10. N.N. Ledentsov, V.A. Shchukin, M. Grundmann, N. Kirstaedter, J. Böhrer, O. Schmidt, D. Bimberg, V.M. Ustinov, A.Y. Egorov, A.E. Zhukov, P.S. Kop'ev, S.V. Zaitsev, N.Y. Gordeev, Z.I. Alferov, A.I. Borovkov, A.O. Kosogov, S.S. Ruvimov, P. Werner, U. Gösele, J. Heydenreich, Direct formation of vertically coupled quantum dots in Stranski-Krastanov growth. Phys. Rev. B **54**, 8743–8750 (1996)
11. L.R.C. Fonseca, J.L. Jiminez, J.P. Leburton, Electronic coupling in InAs/GaAs self-assembled stacked double-quantum-dot systems. Phys. Rev. B **58**, 3643–3648 (1998)
12. H. Ghadi, S. Adhikary, A. Agarwal, J. Agawane, A. Mandal, S. Chakrabarti, One order enhancement of detectivity in quaternary capped InAs/GaAs quantum dot infrared photodetectors due to vertical coupling of quantum dot layers. Thin Solid Films **56**, 1–4 (2014)

Chapter 7
Metamaterial-Based Planar Antennas

Gnanam Gnanagurunathan and Krishnasamy T. Selvan

Abstract Microstrip patch antenna is used extensively in wireless and mobile applications due to its low profile and lightweight. However, this antenna is prone to low gain, limited bandwidth and increased cross polarization levels. Metamaterial can be integrated onto an antenna to improve its performance. A possible approach to enhance the performance is by suppressing surface waves. This can be achieved by using Electromagnetic Bandgap (EBG) structures. In addition, plane waves that come in contact with EBG structures can be reflected in phase thereby enhancing the radiation properties of the microstrip antenna. Therefore, the main motivation underlying this work is to provide an overview on the evolution, characterization and performance enhancement of microstrip antennas with EBG structures.

Keywords Metamaterial · Microstrip patch antenna · Electromagnetic bandgap (EBG) · Artificial magnetic conductor (AMC) · Evolution · Bandgap structure

1 Introduction

"Metamaterials are macroscopic composites having man-made three dimensional, periodic cellular architecture designed to produce an optimized combination, not available in nature, of two or more responses to specific excitation" is aptly described by B.A. Munk in his book [1].

This material is artificially engineered to exhibit a behaviour that is not found in nature. There are many categories of metamaterial that have been established over

G. Gnanagurunathan (✉)
Department of Electrical and Electronic Engineering, The University of Nottingham Malaysia Campus, Jalan Broga, 43500 Semenyih, Selangor Darul Ehsan, Malaysia
e-mail: gnanam.gnanagurunathan@nottingham.edu.my

K.T. Selvan
Department of Electronics and Communication Engineering, SSN College of Engineering, Kalavakkam, Chennai 603110, India

S.R.S. Prabaharan et al. (eds.), *Frontiers in Electronic Technologies*,
Lecture Notes in Electrical Engineering 433, DOI 10.1007/978-981-10-4235-5_7

the years namely, i.e. Double Negative Material (DNG), Single Negative Material (SNG), Electromagnetic Bandgap Structure (EBG), Artificial Magnetic Conductors (AMC), Frequency Selective Sheets (FSS), High Impedance Surfaces (HIS), Chiral media, Zero Index metamaterial (ZIM) and many more.

It should be noted that some of these material may have overlapping properties. In the early years of the periodic structure research, the EBG structure was described as Photonic Bandgap structures whereby the term was developed by the photonic fraternity that worked on stop band performance of optical periodic structures and solid-state electronic band gaps [2]. In [2], the author has also argued that the usage of photonic bandgap and microwave periodic structures (EBG) can be confusing when in actuality both refers to periodic structures that exhibit bandgaps. In this work the focus will be on EBGs and AMCs.

2 Electromagnetic Bandgap (EBG)

Fabricated periodic elements that may prevent or assist or even localize electromagnetic waves propagation within a specified range of frequency [3, 4] can be defined as an EBG structure. These structures are engineered by arranging dielectric material and metallic conductors periodically.

EBG structures can be classified into three types [3]:

- 3D structures—these structures occupy in all three possible direction of the axis and are fabricated and placed in a volumetric form. Examples of these are the woodpile dielectric structure and the multilayer metallic tripod array.
- 2D structures—these are formed on surfaces. There are two variations to these, i.e. mushroom-like EBG which has a via running through the unit cell and uni-planar EBG surface. These types of structures are widely considered in microstrip antenna designs due to its low profile, low cost and in particular ease of fabrication.
- 1D EBG—this is usually in reference to transmission line structures which use a single row of EBG placed along the transmission line.

EBG structures are able to exhibit a unique behaviour depending on whether the incident wave is a plane wave or surface wave. In the case of surface waves that get in contact with EBG, the bandgap of the EBG denies propagation of the surface waves in all angles and polarization states.

On the other hand, if plane waves get in contact with the EBG structures then the reflection phase would vary with frequency. At a particular frequency, the plane wave would encounter a reflection phase which is zero degrees. In reality this would mean the EBG is now functioning as a perfect magnetic conductor (PMC).

3 Evolution

In 1898, Bose [5] carried out experiments of which the outcomes today can be recognized as exhibiting metamaterial properties. His investigations on polarisers made of wired gratings improved the sensitivity of the receivers. He also reported on book-like structures with multiple pages when placed in front of a beam improved the beam's polarization. These were among the earliest documented work on metamaterial. He also documented experimental observation on the twisted fibres of jute that caused an optical twist on the plane of polarization. J. Bose's observation is the very first attempt in observing the microwave behaviour on twisted material, today know as chiral media.

A Finnish scientist, K.F. Lindman, in 1914, explored and investigated artificial chiral media extensively [6]. He managed to configure various models of wired spirals which were used in the study microwaves. His investigation led him to analyzing wave propagation through a grid of wired scatterers. This scientist also studied Frequency Selective Surfaces (FSS) which allow the separation of a range of frequencies from a broad band signal.

In 1946, Leon Brillouin's research left a prominent niche in the EBG world. His breakthrough work on periodic structures and its ability to suppress k (where k is the wavenumber) vectors of waves that propagate within it enabled fundamental and critical understanding of EBG materials [3, 4]. It was this scientist that enabled the determination of the frequency bandgap of a periodic structure using the dispersion curve. His valuable insight on the EBG characterization later became known as the Brilloiun zone.

Experimental investigations on microwave lenses were carried out in 1948, by Kock [7]. The lenses were built using parallel metallic strips. Kock also established an equation to predict the refractive index of the delay lenses based on its structural dimensions. Thereby his observation allowed engineering of artificial material that allowed the lensing effect of microwaves.

In 1968, Veselego [8] hypothesized the possibility of having simultaneous negative values for permittivity (ε) and permeability (μ). In his paper, he also introduced the concept of right handed and left handed substances. Although his paper argues theoretically the possibility to have −ε and −μ, he did put in motion that with technological progress that ε and μ less than zero will be realizable despite identifying limitations on experimental observation.

These scientists pioneering work led to many trail-blazing researches in the field of artificially engineered structures that influences the electromagnetic waves propagation.

In 1987, EBG research took off exponentially due to interesting observations made by two scientists. Yablonovitch [9] analyzed and hypothesized the different aspects of inhibited spontaneous emission whereby if a 3D structure with a band gap overlaps on the electronic band-edge, then it is bound to forbid spontaneous emission. During the same time within the same continent of Canada, John [10] hypothesized a 3D photonic super lattice that can enable strong localization of photons in non-dissipative materials. These observations somewhat influence the definition of the EBG whereby these structures are able to localize or inhibit a range of frequency.

The first periodic structure in an arrangement of dielectric spheres in lattice-like diamond shapes that was easily fabricated and possessed a wide photonic band gap was investigated by Chan et al. [11] in 1991. Within the same year Yablonovitch [12], devised his own periodic structure realized with cylindrical holes that were drilled through the substrate material. This structure was then named Yablonovite and it also exhibited a full photonic band gap.

Today, EBG structures are researched intensively and extensively in the application of microstrip antennas. This is due to its promising properties that allow gain and directivity enhancement, relative bandwidth improvement, size miniaturization and mutual coupling reduction.

4 EBG Properties and Their Characterization

EBG structures are a special class of metamaterial. These structures are able to display various functions such as a Bandgap structure (EBG), Artificial Magnetic Conductor (AMC), High Impedance Surface as well as Soft and Hard surfaces.

This work is focused on the first two functions. Therefore characterizing the EBG as a bandgap structure and as an AMC is a fundamental step towards integrating them on to a patch antenna. The following section describes the EBG's bandgap determination using the dispersion diagram and also obtaining the phase reflection plot to determine the region within which EBG acts as an AMC.

4.1 Bandgap Structure

A typical EBG unit cell is shown in Fig. 1. The bandgap feature of this unit cell can be understood using the lumped LC circuit [3] representation as shown in Fig. 2.

The impedance of a parallel resonant LC circuit is:

$$Z = \frac{j\omega L}{1 - \omega^2 LC} \tag{1}$$

Therefore the resonant frequency of the same circuit is derived as:

$$\omega_0 = \frac{1}{\sqrt{LC}} \tag{2}$$

where Z is the surface impedance and ω_0 is the resonance frequency at which the EBG structure does not support any surface waves.

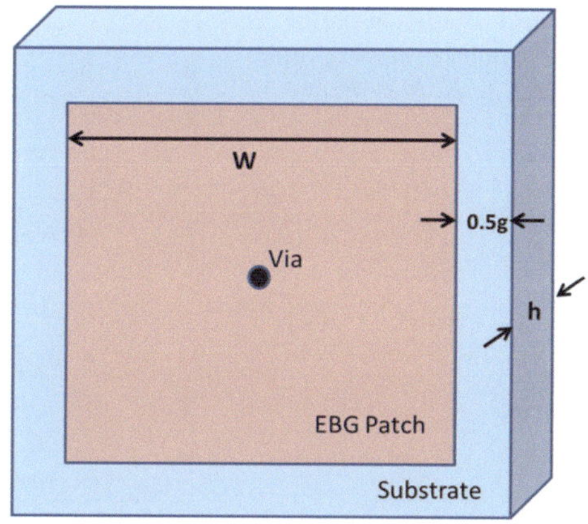

Fig. 1 An EBG unit cell's perspective view and its dimensional parameters

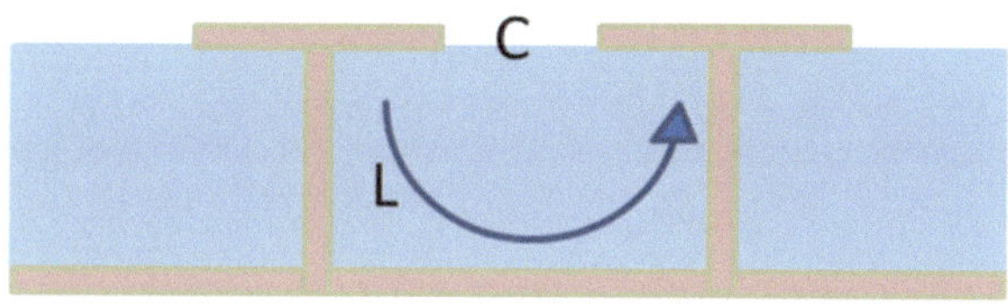

Fig. 2 A cross section view of the unit cells which is used for lumped LC model for EBG analysis

Capacitance, C and inductance L can be determined from the type of material and dimensions that are used:

$$L = \mu h \tag{3}$$

$$C = \frac{W\varepsilon_0(1 + \varepsilon_r)}{\pi} \cos h^{-1}\left(\frac{W + g}{g}\right) \tag{4}$$

where μ is the permeability, ε_r is the dielectric constant, h is the substrate thickness, W is the EBG patch width and g is the width of the gap between the patch.

The approximate computation of the bandgap and impedance is based on static field computations and the fringing fields are not accounted. Defining the Brillouin zone using the dispersion diagram [3] is a more accurate method to determine the surface wave bandgap.

The dispersion diagram is a plot of the frequency versus the wavenumber (k). In order to determine the k for surface waves we need to solve the Eigen-value equation. The solution is not unique to a propagation constant for a particular frequency. Each propagation constant β is known as a specific mode which has its own phase velocity, group velocity and field distribution. Therefore, in a periodic structure the surface wave is also periodic in nature with a phase delay dependent on the wavenumber k and periodicity p.

The dispersion diagram is plotted for a single period for a two dimensional square area where $0 \le k_{xn} \le \frac{2\pi}{p_x}$ and $0 \le k_{yn} \le \frac{2\pi}{p_y}$.

Table 1 Variation on the wavenumbers; moving from point Γ to X, X to M and M back to Γ

Γ: ⇩ X:	$k_x = 0$ ⇩ $k_x = \frac{2\pi}{(W+g)}$ (k_x *varied*)	$k_y = 0$ (k_y *remains constant*)
X: ⇩ M	$k_x = \frac{2\pi}{(W+g)}$ (k_x *remains constant*)	$k_y = 0$ ⇩ $k_y = \frac{2\pi}{(W+g)}$ (k_y *varied*)
M ⇩ Γ:	$k_x = \frac{2\pi}{(W+g)}$ ⇩ $k_x = 0$ (k_x *varied*))	$k_y = \frac{2\pi}{(W+g)}$ ⇩ $k_y = 0$ (k_y *varied*)

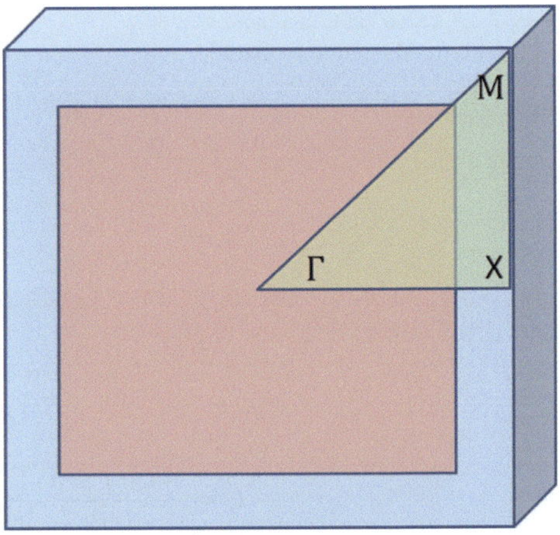

Fig. 3 The k diagram

The dispersion diagram plot's vertical axis illustrates the frequency and horizontal axis illustrates the transverse wavenumbers (k_x and k_y) over three specific points, i.e. Γ, X and M. Moving from point Γ to X, X to M and M back to Γ, the wavenumbers are varied as shown in Table 1 and the resonant frequencies of the surface modes are identified.

Figure 3 illustrates the EBG unit cell and the Γ, X and M constellation over which the wave numbers are determined. Due to the geometry's symmetry, the triangular Brillouin zone is applicable over the entire unit cell.

4.2 Transmission Line Method to Determine the Bandgap

The transmission line method is an alternative method to determine the EBG structure's bandgap. A periodic lattice and a transmission line sandwiching the substrate can be assumed as a 2 port network. Based on the S-parameters observed upon the 2 ports, an attenuation of S_{21} observed over a range of frequency will be the intended bandgap [13]. This similar approach is also used for filter designs.

The effect of introducing etched square and circular pattern on the EBG bandgap was analyzed in [14]. In [15] a two layer structure is analyzed for its bandgap. This approach allows mushroom-like structures to be characterized. Much research has been undertaken along this transmission line method to identify the bandgap of an EBG structure [16–21].

4.3 Artificial Magnetic Conductor (AMC)

EBG structures in addition to suppressing surface waves, can also act as an artificial magnetic conductor. As the name implies this kind of structure is not available in nature, however, it can be engineered. Normally, an incidental plane wave on a perfect electric conductor (PEC) has a reflection co-efficient of −1, in other words the reflected wave is 180° out of phase. In the case of an EBG structure the reflected phase is 0°, thereby being in phase with the incidental wave [3]. These EBG structures that function as Perfect Magnetic Conductor (PMC) are known as AMC.

The reflection phase of an EBG is a function of frequency. The phase of reflection is made to vary from −180° to +180° as the frequency is made to increase. For the purpose of AMC characterization, the scattered fields from an ideal PEC and an EBG surface are observed. Normalization of the reflected phase from the EBG structure to the reflected phase of the PEC surface is then performed. Subsequently, a π factor is then added to the phase result. This allows to account for the reference of a PEC surface [3, 22]. This characterization enables the determination of the frequency at which the EBG structure functions as an AMC. Usually the frequency band over which the EBG structure is able to exhibit AMC characteristics is defined for a phase reflection of −90° to +90° [23].

The EBG's property as an AMC is capitalized by using it as ground plane to improve the radiation characteristics of a microstrip antenna.

5 EBG Integrated Patch Antennas

Extensive research on patch antennas integrated with EBG structures are being carried out. Due to the dynamic status of this research field, the scope of this descriptive survey in this thesis is limited to microstrip antennas and the improvements that have been achieved in their performance using EBG structures.

In general, the following section will describe on the improvements that can be achieved using EBG structures on microstrip antennas, i.e.

- Improving the radiation characteristics
- Achieving miniaturization
- Eliminating mutual coupling
- Functioning as a filter

5.1 Improving the Radiation Characteristics

EBG structures can be integrated onto an antenna in any one of the following manner to improve the radiation characteristics:

- surrounding the radiating patch
- below the radiating patch as a substrate or as a ground plane
- above the radiating patch as a superstrate

The above are described hereafter.

5.1.1 EBG Positioned Around the Radiating Patch

Surrounding the patch antenna with an EBG structure suppresses the surface waves. Surface waves are prominent in thick substrates which are usually used to improve the bandwidth of a microstrip antenna. Suppressing surface waves in the substrate allows an increased amount of radiated power to couple to space waves thereby enhancing the gain of an antenna [24–26]. In addition, this can also mitigate diffracted surface waves at the edge of the patch that can aggravate back lobe radiation [27]. Power losses which are prominent in thick substrates and high dielectric constant can also be reduced [27].

Therefore suppressing surface waves by enclosing a radiating patch with an EBG structure is one possible approach to further enhance the performance of a patch antenna. A planar fed antenna surrounded by a uniplanar compact EBG (UCEBG) structure was investigated in [24]. It reported on a UC-EBG cell's dispersion diagram analysis and subsequent implementation of the EBG surrounding a 12 GHz patch antenna. The computation and measurement results indicated a more focused radiation pattern and also a gain improvement of almost 3 dB.

A patch antenna placed on an array of air–columns embedded on a 10 mm dielectric substrate functioning as an EBG structure was investigated in [27]. This investigation reported improved gain by 10 dB and improved radiation pattern whereby the back lobe radiation is reduced. However, it should be noted that this researcher carried out his investigation on a 10 mm thick substrate to ensure significant presence of surface waves.

A comparison between a step like substrate structure and mushroom-like EBG consisting substrate was investigated in [28] on a high dielectric constant substrate. The radiation performance of both antenna structures improved significantly. In addition the back lobe radiation was also minimized. Similar observation has also been reported by other researchers [29].

Investigations on these type of antennas with enclosed EBG structures was also investigated for dual-band antennas [30] and antennas in the terahertz frequency [31]. A dual-band antenna comprising of two radiating patches stacked one onto the other and each patch surrounded by fractal EBG has enabled low profile structure for dual-band applications. The gain was improved in addition to wider impedance bandwidth [30]. EBG structures have also showed improved directivity for antennas in the terahertz frequency. However, this investigation was carried using simulation and experimental verification was scaled down. Despite this both simulation and measurement indicated improved directivity for the planar fed patch antenna at terahertz frequency [31].

Drawback of this approach is that the gain improvement observed for this kind of implementation is very low, about 3 dB except in [27]. However, radiation pattern improvements reported were remarkable. Antenna with these EBG structures also have increased dimension compared to conventional patch antennas.

5.1.2 EBG Positioned Below the Radiating Patch

Another possibility of integrating the EBG structure on a patch antenna is below the radiating patch sandwiched between the grounds. In this position the EBG functions as reflector surface and usually known as an AMC. Various efforts have been reported using AMC integrated on a patch antenna and various improvements have been reported in respect of backlobe reduction, gain and impedance bandwidth improvement.

A novel UC-EBG structure functioning as an AMC was investigated in [32] and experimentally verified. The researcher reasons that one way to differentiate the PEC and PMC is by looking at the surface impedance. The surface impedance of a PMC is an open circuit and therefore a periodic pattern would be able to create an open circuit condition.

AMCs sandwiched on the ground plane of a probe fed patch antenna [33], plane fed patch antenna [34] and a plane fed wideband [35–37] antenna have reportedly improved the gain and the impedance bandwidth. In [33], a detailed parametric analysis was carried out on a two layer microstrip patch antenna working within the bandgap and outside the bandgap. Outside the bandgap, 25% improvement on bandwidth and 3 dB gain improvement were obtained. Within the bandgap, impedance bandwidth improved 20% whereas gain improved by 10 dB. The implementation of an AMC on a wideband antenna [35] shows good gain improvement and wider impedance bandwidth.

In [38] a multi periodic EBG structure sandwiched between the ground planes was analyzed. Comparisons were made against a periodic EBG structure. There was a slight improvement in the gain of the multi periodic EBG structure by 0.6dBi. A slight improvement in the impedance bandwidth was also noted.

Investigations in [39] reported on suppressing the parallel-plate modes that exist in an aperture coupled antenna. By introducing an AMC structure, back lobe suppression was achieved. Two types of reflector surfaces formed by an AMC and a PEC were compared against the antenna without the reflectors. It was reported that the sidelobe suppression was significant for both the PEC and AMC structures compared to the antenna without reflectors. The PEC and AMC surfaces, however, showed similar back lobe suppression.

Dual-band characteristic using a single radiating element was also reported using 2 layers of EBG substrates [40]. However, it was reported additional unwanted resonant frequencies appeared due to parasitic coupling between radiating patch and the EBG structure during measurements.

Wearable antennas have also benefitted from EBG placed below the radiating patch [41]. It is reported that the EBG structure has reduced the radiation into the human body by over 15 dB and the effect of frequency detuning caused by the human body. This was made possible due to the back lobe suppression made possible by the EBG placement.

5.1.3 EBG Positioned Above the Radiating Patch

EBG structures can also be placed above the radiating patch. By doing so, the radiation characteristic of a patch antenna can be improved. Various investigations have been carried out to illustrate this improvement on patch antennas and also on wideband antennas. The superstrate structures are also known as Frequency Selective Sheet (FSS).

Square loop [42] and circular ring [43] FSS sheets have been used to enhance the directivity of a patch antenna. In addition wideband and broad band antennas too have seen improvement in gain by using EBG structures as superstrate [44, 45]. This is crucial as gain is usually compromised in wideband and dual-band structures in comparison to single resonant antenna.

Studies have also been carried out using genetic algorithm to enable optimized design of the FSS [46] which showed improved gain and directivity. In addition, in [47], the researcher has used several radiating elements as in an array antenna to excite the EBG structure. This approach improved the radiation bandwidth and the gain by almost 2 dB.

Most of these works were carried out with a maximum of two layers of EBG structure. This is to ensure that the dimensional increase in the patch antenna's profile is proportional with its radiation characteristic improvement.

5.2 Enabling Miniaturization

Achieving miniaturization for wireless antennas are crucial due to ever evolving demands for more compact and low profile communicating devices. EBGs can be integrated onto a patch antenna to achieve miniaturization

In [48] a patch antenna designed to resonate at 10 GHz was integrated with an EBG structure on the ground plane. This antenna configuration exhibited a resonant frequency of 2.4 GHz thereby enabling a size reduction of 75%. In addition the radiation pattern was acceptable with a gain value of 2.4 dB.

The investigation in [49] compared the performance of the conventional mushroom-like EBG with a spiral EBG. The spiral structure showed a drop in the bandgap frequency from 6 to 7 GHz to almost 2–3 GHz for the same dimension of a unit cell. Therefore miniaturization is achieved.

The performance of a miniaturized antenna designed with meandered slots for Bluetooth application was further improved with the usage of an EBG structure [50]. The EBG structure placed on the ground plane increased the transceiver's distance significantly by 22 m.

In addition, miniaturization was achieved in [51] by placing an AMC as a reflector for a slotted patch antenna which serves as a telemetry system using the 2.45 GHz band to enable signal transmission and reception in indoors and the 6–7.5 GHz band for radar sensing part. The radar sensing was made possible due to the AMC reflector working as a perfect magnetic conductor (PMC) plane from 6 to 7.5 GHz.

5.3 Mitigating Mutual Coupling

Array antennas are prone to mutual coupling and this issue is usually solved using complex feed networks. EBG structures have been recognized to mitigate the mutual coupling between array antennas which is a simpler alternative to complex feed networks. By reducing the mutual coupling, the directivity and gain of a microstrip antenna can be improved.

In [52], it was demonstrated that the mushroom-like EBG embedded between two radiating patches could substantially reduce mutual coupling by up to 8 dB.

Similar mutual coupling reducing works were also carried out by other researchers. The research of [53] was implemented on two slot antennas on a single substrate which improved the radiation pattern and the gain.

On the other hand, the investigation in [54] was done on a 2 layer structure whereby the EBG structure was placed on top of a dual radiating element. The EBG on the top layer forms a row of cells which appears between the radiating elements. This study showed a 10 dB reduction in mutual coupling.

5.4 Performing as a Filter

The EBG's property of exhibiting a bandgap over a range of frequencies is exploited to function as a bandstop filter in microstrip antenna applications. The research carried out in [55] implemented a single row of EBG cells on the ground plane placed directly below in parallel to the planar feed. These EBG cells act as a filter to inhibit a desired band of frequencies between 3–4 GHz, while the impedance bandwidth of the antenna without the EBG filter is from 2 to 6 GHz.

Similar integration of the EBG as a filter was also implemented on an array antenna [56]. These structures were located under the feed line and in addition a combination of EBG structures are analyzed:

- Mushroom-like EBG structure
- Minkowski and mushroom-like EBG structure
- Sierpinski and mushroom-like EBG structure

The radiation pattern improved for all the combinational structures compared to the mushroom-like structure only.

6 Summary

An overview on the evolution of the metamaterial generally and EBG more specifically is covered in this paper. In addition, various possible EBG structure characterization methods, possible integration onto a patch antenna and its ability to improve the radiation characteristics were also reviewed. This overview lays the foundation for researchers to further explore the myriads of applications that can benefit from the integration of metamaterial on to antennas.

Acknowledgements This paper is based on the primary author's doctoral write-up [57].

References

1. B.A. Munk, *Metamaterials: Critique and Alternatives* (Wiley, Hoboken, 2009)
2. A.A. Oliner, Periodic structures and photonic-band-gap terminology: historical perspectives, in *29th European Microwave Conference, 1999*, (Munich, Germany, 1999), pp. 295–298
3. F. Yang, Y. Rahmat-Samii, *Electromagnetic Bandgap Structures in Antenna Engineering* (Cambridge University Press, 2009)
4. N. Engheta, R.W. Ziolkowski, *Metamaterials: Physics and Engineering Explorations* (John Wiley & Sons, Inc, 2006)
5. J.C. Bose, On the rotation of plane of polarisation of electric waves by a twisted structure. Proc. R. Soc. Lond. **63**, 146–152 (1898)
6. I. Lindell, A.H. Sihvola, J. Kurkijarvi, Karl f Lindman: the last hertzian and a harbinger of electromagnetic chirality. IEEE Antennas Propag. Mag. **34**(3), 24–30 (1992)

7. W.E. Kock, Metallic delay lenses. Nature **163**, 324–325 (1949)
8. V.G. Veselago, The electrodynamics of substances with simultaneously negative values of E and μ. Sov. Phys. Uspekh **10**, 509 (1968)
9. E. Yablonovitch, Inhibited spontaneous emission in solid-state physics and electronics. Phys. Rev. Lett. **58**, 2059–2062 (1987)
10. S. John, Strong localization of photons in certain disordered dielectric superlattices. Phys. Rev. Lett. **58**, 2486–2489 (1987)
11. C.T. Chan, K.M. Ho, C.M. Soukoulis, Photonic band gaps in experimentally realizable periodic dielectric structures. Europhys. Lett. **16**, 563 (1991)
12. E. Yablonovitch, T.J. Gmitter, Photonic band structure: the face-centered-cubic case employing nonspherical atoms. Phys. Rev. Lett. **67**, 2295–2298 (1991)
13. X. Ying, A. Alphones, Propagation characteristics of complimentary split ring resonator (CSRR) based EBG structure. Microw. Opt. Technol. Lett. **47** (2005)
14. M.M. Karbassian, H. Ghafouri-Shiraz, Effect of shape of patterns on the performance of microstrip photonic band-gap filters. Microw. Opt. Technol. Lett. **48**, 1007–1011 (2006)
15. N. Yang, Z.N. Chen, Y.Y. Wang, M.Y.W. Chia, A two-layer compact electromagnetic bandgap (EBG) structure and its applications in microstrip filter design. Microw. Opt. Technol. Lett. **37** (2002)
16. S.K. Menon, K. Vasudevan, C.K. Aanandan, P. Mohanan, Design and analysis of microstrip lines with EBG-backed ground planes of different geometrical shapes. Microw. Opt. Technol. Lett. **46**, 544–546 (2005)
17. S.M. Moghadasi, Compact and Wideband 1-D mushroom-like EBG filters. Prog. Electromagnet. Res. **83**, 323–333 (2008)
18. B.Q. Lin, X.-Y. Ye, X.-Y. Cao, F. Li, Uniplanar EBG structure with improved compact and wideband characteristics. Electron. Lett. **44**, 1362–1363 (2008)
19. J.D. Ruiz, F.L. Martinez, J. Hinojosa, 1D Koch fractal electromagnetic bandgap microstrip structures with r/a ratios higher than 0.5. Microw. Opt. Technol. Lett. **53**, 646–649 (2011)
20. S.K. Padhi, Improved performance of EBGs on a co-planar transmission line using tapered distribution. Microw. Opt. Technol. Lett. **42**, 128–131 (2004)
21. G. Gnanagurunathan, K.T. Selvan, Performance analysis of complementary and non-complementary EBG geometries, presented at the progress, in *Electromagnetics Research Symposium (PIERS 2012)*, (Kuala Lumpur, Malaysia, 2012)
22. F. Yang, Y. Rahmat-Samii, Reflection phase characterizations of the EBG ground plane for low profile wire antenna applications. IEEE Trans. Antennas Propag. **51**, 2691–2703 (2003)
23. D. Sievenpiper, L. Zhang, R.F.J. Broas, N.G. Alexopolous, E. Yablonovitch, High-impedance electromagnetic surfaces with a forbidden frequency band. IEEE Trans. Microw. Theory Tech. **47**, 2059–2074 (1999)
24. R. Coccioli, F.-R. Yang, K.-P. Ma, T. Itoh, Aperture-coupled patch antenna on UC-PBG substrate. IEEE Trans. Microw. Theory Tech. **47**, 2123–2130 (1999)
25. Y. Qian, R. Coccioli, D. Sievenpiper, V. Radisic, E. Yablonovitch, T. Itoh, Microstrip patch antenna using novel photonic band-gap structures. Microw. J. **42**, 6676 (1999)
26. G. Gnanagurunathan, K.T. Selvan, Gain enhancement of microstrip patch antenna by using complementary EBG geometries. J. Electromagnet. Waves Appl. **26**, 329–341, (2012) (2012/01/01)
27. R. Gonzalo, P. Maagt, M. Sorolla, Enhanced patch-antenna performance by suppressing surface waves using photonic-bandgap substrates. IEEE Trans. Microw. Theory Tech. **47**, 2131–2138 (1999)
28. F. Yang, C.-S. Kee, Y. Rahmat-Samii, Step-like structure and EBG structure to improve the performance of patch antennas on high dielectric substrate, in *IEEE Antennas and Propagation Society International Symposium*, (Boston, 2001), pp. 482–485
29. M. Fallah-Rad, L. Shafai, Enhanced performance of a microstrip patch antenna using a high impedance EBG structure, in *IEEE Antennas and Propagation Society International Symposium, 2003,* (2003), pp. 982–985

30. X.L. Bao, G. Ruvio, M.J. Ammann, Low-profile dual-frequency GPS patch antenna enhanced with dual-band EBG structure. Microw. Opt. Technol. Lett. **49** (2007)
31. K.R. Jha, G. Singh, Analysis and design of enhanced directivity microstrip antenna at terahertz frequency by using electromagnetic bandgap material, in *International Journal of Numerical Modelling: Electronic Networks, Devices and Fields,* vol. 24 (2010)
32. K.P. Ma, K. Hirose, F.-R. Yang, Y. Qian, T. Itoh, Realisation of magnetic conducting surface using novel photonic bandgap structure. Electron. Lett. **34**, 2041–2042 (1998)
33. D. Qu, L. Shafai, A. Foroozesh, Improving microstrip patch antenna performance using EBG substrates. IEE Proc. Microw. Antennas Propag. **153**, 558–563 (2006)
34. C.C. Chiau, X. Chen, C. Parini, Multiperiod EBG structure for wide stopband circuits. IEE Proc. Microw. Antennas Propag. **150**, 489–492 (2003)
35. D.N. Elsheak, M.F. Iskander, H.A. Elsade, E.A. Abdallah, H. Elhenawy, Enhancement of ultra-wideband microstrip monopole antenna by using unequal arms V-shaped slot printed on metamaterial surface. Microw. Opt. Technol. Lett. **52**, 2203–2209 (2010)
36. G. Gnanagurunathan, K.T. Selvan, Artificial magnetic conductors on wideband patch antenna. Progress Electromagn. Res. Lett. **36**, 9–19 (2013)
37. W. Yang, H. Wang, W. Che, J. Wang, A wideband and high-gain edge-fed patch antenna and array using artificial magnetic conductor structures. IEEE Antennas Wirel. Propag. Lett. **12**, 769–772 (2013)
38. C.C. Chiau et al., A sandwiched multiperiod EBG structure for microstrip patch antennas. Microw. Opt. Technol. Lett. **46**, 437–440 (2005)
39. Y. Zhang, J. von Hagen, M. Younis, C. Fischer, W. Wiesbeck, Planar artificial magnetic conductors and patch antennas. IEEE Trans. Antennas Propag. **51** (2003)
40. X. J. Wang, Y. Hao, Dual-band operation of an electromagnetic band-gap patch antenna. Microw. Opt. Technol. Lett. **49** (2007)
41. S. Velan, E.F. Sundarsingh, M. Kanagasabai, A.K. Sarma, C. Raviteja, R. Sivasamy et al., Dual-band EBG integrated monopole antenna deploying fractal geometry for wearable applications. IEEE Antennas Wirel. Propag. Lett. **14**, 249–252 (2015)
42. A. Pirhadi, F. Keshmiri, M. Hakkak, M. Tayarani, Analysis and design of dual band high directive EBG resonator antenna using square loop FSS as superstrate layer. Progress Electromag. Res. **70**, 1–20 (2007)
43. D.H. Lee, Y.J. Lee, J. Yeo, R. Mittra, W.S. Park, Directivity enhancement of circular polarized patch antenna using ring-shaped frequency selective surface superstrate. Microw. Opt. Technol. Lett. **49**, 199–201 (2007)
44. Z.-C. Ge, W.-X. Zhang, Z.-G. Liu, Y.Y. Gu, Broadband and high-gain printed antennas constructed from Fabry–Perot resonator structure using EBG or FSS cover. Microw. Opt. Technol. Lett. **48**, 1272–1274 (2005)
45. L. Moustafa, B. Jecko, Design of a wideband highly directive EBG antenna using double-layer frequency selective surfaces and multifeed technique for application in the ku-band. IEEE Antennas Wirel. Propag. Lett. **9**, 342–346 (2010)
46. Y. Ge, K.P. Esselle, Y. Hao, Design of low-profile high-gain EBG resonator antennas using a genetic algorithm. IEEE Antennas Wirel. Propag. Lett. **6**, 480–483 (2007)
47. L. Leger, T. Monediere, J. Bernard, Enhancement of gain and radiation bandwidth for a planar 1-D EBG antenna. IEEE Microw. Wirel. Compon. Lett. **15**, 573–575 (2005)
48. A.A. Eldek, A miniaturized patch antenna at 2.4 GHz using uni-planar compact photonic band gap structure. Microw. Opt. Technol. Lett. **50**, 1360–1363 (2008)
49. H.-H. Xie, Y.-C. Jiao, K. Song, B. Yang, Miniature electromagnetic band-gap structure using spiral ground plane. Progress Electromag. Res. Lett. **17**, 163–170 (2010)
50. M.F. Karim, H. Ghafouri-Shiraz, EBG-assisted slot antenna for Bluetooth applications. Microw. Opt. Technol. Lett. **48**, 482–487 (2006)
51. S. Yan, P.J. Soh, M. Mercuri, D.M.M.P. Schreurs, G.A.E. Vandenbosch, Low profile dual-band antenna loaded with artificial magnetic conductor for indoor radar systems. IET Radar Sonar Navig. **9**, 184–190 (2015)

52. F. Yang, Y. Rahmat-Samii, Microstrip antennas integrated with electromagnetic band-gap (EBG) structures: a low mutual coupling design for array applications. **51**, 2936–2946 (2003)
53. K. Payandehjoo, R. Abhari, Employing EBG structures in multiantenna systems for improving isolation and diversity gain. IEEE Antennas Wirel. Propag. Lett. **8**, 1162–1165 (2009)
54. H.S. Farahani, M. Veysi, M. Kamyab, A. Tadjalli, Mutual coupling reduction in patch antenna arrays using UC-EBG superstrate. IEEE Antennas Wirel. Propag. Lett. **9**, 57–59 (2010)
55. F. Consoli, R. Catalano, R. Laudani, L. Tumino, S. Barbarino, Planar slot antenna with PBG filter for wireless communications. Microw. Opt. Technol. Lett. **49**, 551–555 (2007)
56. T. Masri, M.K.A. Rahim, Dual-band microstrip antenna array with a combination of mushroom, modified Minkowski and Sierpinski electromagnetic band gap structures. IET Microw. Antennas Propag. **4**, 1756–1763 (2010)
57. G. Gnanagurunathan, *Electromagnetic Bandgap stucture based patch antenna* (PHD, Department of Electrical and Electronic Engineering, University of Nottingham, 2012)

Chapter 8
An Overview on Memristor-Based Non-volatile LUT of an FPGA

T. Nandha Kumar

Abstract This paper presents detailed over view of four different novel memristor-based nano-crossbar structures that are employed for designing a Non-Volatile Look-Up Table (NVLUT) of a Field Programmable Gate Array (FPGA) with particular focus on efficient READ and WRITE operations; techniques to overcome the effect sneak path current on unselected memristors. Cross-point memory array structure, MOS accessed memory array structure, Memory array structure with transistors at BL and Columns isolated memory array structure are the different structures considered in this paper. The paper analyzes in detail about the LUT structures and their WRITE and READ operations. Also the analysis on the SPICE simulation results of the READ and WRITE delay, energy dissipation and energy delay product of four LUT structures are presented. Among the LUT structures, Columns isolated memory array structure seems to be potential candidate for NVLUT as it eliminates the effect of sneak path current on the unselected memristors. Also it showed a fast WRITE time, significantly reduced READ power dissipation and no power dissipation in the stand-by mode. In addition, it eliminates the write half-select issue. Moreover, it prevents the data integrity when compared with other structures as this structure has better controllability of the NSPs for the unselected memristors.

Keywords Memristor · Non-volatile memory · FPGA · LUT

1 Introduction

Field Programmable Gate Arrays (FPGAs) when compared with Application Specific Integrated Circuit (ASIC) design helps in hardware realization of digital design at a rapid rate and relatively low development cost [1]. The common Static

T. Nandha Kumar (✉)
Department of Electrical and Electronics Engineering,
University of Nottingham, Nottingham, Malaysia
e-mail: nandhakumaar.t@nottingham.edu.my

S.R.S. Prabaharan et al. (eds.), *Frontiers in Electronic Technologies*,
Lecture Notes in Electrical Engineering 433, DOI 10.1007/978-981-10-4235-5_8

Random Access Memory cell (SRAM) FPGA architecture consists of a regular, flexible and programmable two-dimensional array of Configurable Logic Blocks (CLBs) [2]. Usually, a CLB consists of Look-Up Tables (LUTs), multiplexers and Flip-Flops (FFs) [1]. The combinational part of the logic design is implemented in LUTs. All configurable resources in the FPGA (inclusive of the LUTs) are controlled by the configuration bits stored in a SRAM [2]. However, with a malfunction occur at the power supply or power supply is interrupted then a SRAM is unable to retain the configurations bits leading to a danger of losing or corrupting the actual hardware design. Thus as a possible solution, non-volatile NAND flash memory is used in storing the configuration bits [3], therefore it is integrated into the FPGA [4]. However this solution leads to issues such as a larger silicon area, increase in cost and more importantly very slow data retrieving time. In addition as technology enters the very deep submicron and nano scales, a substantial increase of leakage current is encountered when the FPGA is in stand-by mode, thus causing additional power dissipation [5].

Hence to overcome the above mentioned issues, non-volatile (NV) memory blocks (as LUTs) made of resistive elements such as the memristor has drawn a great interest among the researchers. The memristor is the fourth fundamental passive element postulated by Chua [6] and realized using a nano scale thin film of titanium di-oxide sandwiched between two platinum electrodes by scientists at Hewlett-Packard (HP) Labs [7].

Memristor-based memories using nano-crossbars have been extensively analyzed in the technical literature [5, 8–24]. These memories have been promoted as a potential replacement for conventional NV Flash Memories (NVFM) as LUTs in a FPGA due to the higher density and lower power consumption [8]. Also memristors can outperform many other emerging non-volatile memories (NVMs) in features such as area, delay and power dissipation [8].

A NV LUT in a FPGA must meet some requirements that make a memristor-based implementation very attractive: its size is usually small and once programming (WRITE) for an application is accomplished, an FPGA requires fast READ operation [5]. In addition, multiple READ operations should not affect the memristance of the memristor. Thus the logic state of the memristor needs to be preserved. Usually NV LUT design comprises of the memristors connected onto a nano-crossbar, in such case the leakage or sneak path current should not affect the memristance of the memristor that are not targeted for operation.

To address the above issues researchers have developed different nano-crossbar-based memory structures. The crossbar structure known as Cross-point memory array that has a NMOS pass transistor on each word line (WL) and bit line (BL) is proposed by [10, 24]. Next MOS accessed memory array structure that has a NMOS pass transistor connected to each memristor at each cross point of the nano-crossbar is proposed in [11]. This structure uses 0.5 Vdd on the unselected WL and BL. The next structure that uses only NMOS pass transistors on the BL is proposed by [5, 15, 17]. It utilizes new WRITE and READ operations. This structure requires a RESTORE pulse only for the READ 0 operation, thus it saves 25% of the READ time when compared with previous methods [10, 11, 24].

As for the power rails, it also does not require a 0.5 Vdd power rail, hence it reduces the number of power rails of [11] by 1 and requires no additional circuit to handle the WRITE 1 (0) operation for both WL and BL. Simulation and comparative analysis of this design with other memristor-based memories [10, 11, 24] has shown that it is significantly better in terms of both WRITE and READ times as well as energy consumption. However, the above mentioned three structures suffer from the sneak path current that affects the untargeted memristors in the crossbar. To address these issue, [16, 18] propose a novel structure in which the memristors are connected in rows and columns, but the columns are isolated. It is then possible to prevent the sneak path current and write half-select problems, because the memristances of the unselected memristors are unaffected. Also this structure retains the advantages of previous structures, such as no power dissipation in stand-by mode; moreover, no refresh pulse and no 0.5 Vdd bias are required. Also lowering the number of power rails.

This paper presents detailed overview on above mentioned different memristor-based nano-crossbar structures that are used for designing the NV LUT for FPGA. Particular focus is on efficient READ and WRITE operations; techniques to overcome the sneak path current. The rest of the paper is organized as follows: Sect. 2 deals with the preliminaries and review of this work. The proposed different NV LUT structures and their operations are presented in Sect. 3. Section 4 presents discussion on the simulation results of different NV LUT structures. Section 5 concludes this manuscript.

2 Preliminaries and Review

The implementation of the memristor by HP has two layers of a thin film of titanium di-oxide, one is doped with a high concentration of positive ions, while other is slightly doped [7]. The charge carriers are oxygen vacancies, i.e. the doped layer has a significantly lower resistance than the non-doped layer. The resistance of the film is dependent on the charge that is passed through it in a particular direction; this process can be reversed by changing the direction of the current [7].

Let w(t) denote the length of the doped region (as function of time) and D the total length of the titanium dioxide layer (memristor); the off and on resistances are the extreme values of the memristor resistance. The range of memristor operation is defined in terms of the location of the TiO_2 boundary [7]. The conductivity of a memristor depends on the length of the doped region; if a sufficiently high charge/flux is passed through the memristor, then the oxygen ions spread over the entire film, the resistance decreases and the memristor is said to be in the ON state (with resistance given by R_{ON}). The application of a different polarity voltage can reverse this phenomenon and if no doped region is present, then the memristor is in the OFF state with high impedance (R_{OFF}) [7]. The values over which the resistance varies define the so-called *range of the memristor*, i.e. larger the film length or lower the conductivity of the non-doped region is, greater is the range. As stated in [5, 15–18] the following definition is applicable to this paper as also used throughout the technical literature.

Definition 1 w(t)/D is referred to as the Normalized State Parameter (NSP).

When w(t) = D, then NSP = 1 and the memristor is at the least resistance value (R_{ON}). If w(t) = 0 then, NSP = 0 and the memristor is at the highest resistance value (R_{OFF}). The threshold value of the NSP is the value that allows correctly distinguishing the binary values (0, 1) of the stored data; in this paper, this is given by 0.5.

Recent NV memory researches have resulted in novel FPGA architectures [8, 25–28]. In [28] a 3-D architecture to partition the transistors in the interconnect and logic into active layers and stacks them using a monolithic stacking process is employed. A novel architecture instead of using 3-D stacking process uses only memristors and metal wires for implementing the interconnect of a FPGA has been proposed in [8]. Drowsy mode and body biasing techniques are used in [29, 30] to reduce the power consumption of a FPGA. Turkyilmaz et al. [31] uses Non-Volatile SRAM (NVSRAM) memories with bipolar OxRRAM to reduce power at stand-by mode. In all the above mentioned methods, NV operation is obtained by utilizing a SRAM cell for storage, while adding memristors to it for NV operation. Hence, such methods suffer from dynamic power dissipation (due to the additional number of transistors) and incur in substantial area and delay overheads. Moreover, WRITE and READ operations of such methods are complicated once a memristor is integrated in the memory cell.

3 NV LUT Structures and Operations

New development in NVMs has resulted in many novel memristor-based structures for FPGA architectures. The NV LUT for a FPGA comprises of nano-crossbar with its every cross junction connected with a memristor. Such structures have to ensure a memristor-based implementation very attractive with regards to size, fast WRITE and READ operations, retaining NSP of memristor during READ operations and due to the sneak path current. The four different memristor-based nano-crossbar structures to be discussed in this paper are namely (i) *Cross-point memory array structure*, (ii) *MOS accessed memory array structure*, (iii) *Memory array structure with transistors at BL* and (iv) *Columns isolated memory array structure.*

(i) *Cross-point memory array structure* [10, 24]

This structure employs a memristor at every junction of the nano-crossbar. In addition, each BL and WL are connected to a NMOS pass transistor. A 2 × 2 LUT structure is shown in Fig. 1 where M11, M12, M21 and M22 are memristors connected at the junctions of the crossbar.

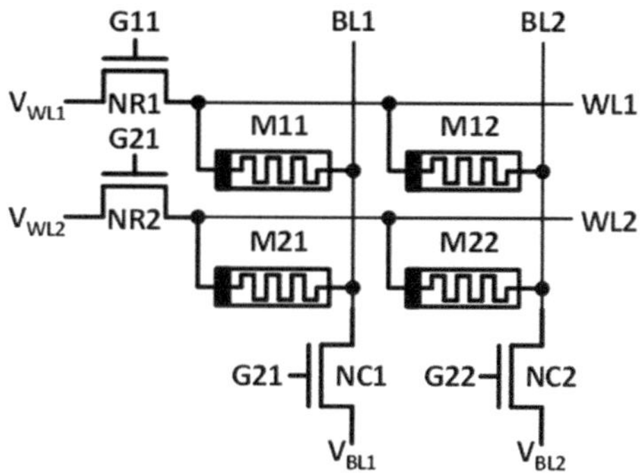

Fig. 1 Two-input LUT using cross-point memory array structure

In this method, to select a memristor (M11) for a WRITE or READ operation requires the transistor (NR1) in the corresponding word line (WL1) and the transistor (NC1) in the corresponding bit line (BL1) to be switched ON. To unselect the other memristors, the remaining word line (WL2) and bit line (BL2) are connected to 0.5 Vdd. The WRITE operation is then performed on M11 as follows. (a) For WRITE 1, +Vdd is applied to WL1 and BL1 is connected to ground. (b) For WRITE 0, +Vdd is applied to BL1 and WL1 is grounded. Thus, an additional circuit is required to identify the value to be written and accordingly apply +Vdd to either WL or BL of the memory cell to which the corresponding memristor is connected. Moreover, the requirement of using +Vdd and 0.5 Vdd causes a voltage difference across an unselected memristor and hence, its NSP change.

The READ operation consists of a READ pulse followed by a RESTORE pulse. The READ operation is performed by applying a short duration negative pulse at WL1 of the selected memristor M11; a RESTORE pulse (a short duration positive pulse) is applied to the word line of the selected memristor after every READ operation. When performing a READ 1, the READ pulse reduces the NSP of the memristor because the READ pulse has a negative magnitude (−Vdd); therefore, a positive pulse is applied to restore its NSP to the original value. However, the NSP may not reach its original value due to an insufficient width of the RESTORE pulse. Thus, consecutive READ operations may change the NSP of the memristor and after a few READ 1 operations, a REFRESH pulse must be applied to ensure the NSP does not cross the threshold value. In addition, this method uses four power rails (+Vdd, −Vdd, 0.5 Vdd and GND). The WRITE and READ voltages are presented in Table 1.

(ii) *MOS accessed memory array structure* [11]

This structure similar to cross-point memory structure employs a nano-crossbar and its junctions memristors are connected. However, different from [10, 24], each connected memristor is isolated using MOS access device as shown in Fig. 2. In addition, apart from WL and BL, select line (SL) is used for WRITE and READ operation, while WL is used for selecting the appropriate memristor to READ or WRITE operation.

Table 1 Voltage requirement for WRITE and READ operations on selected memristor (M11) using (i) Cross-point memory array structure, (ii) MOS accessed memory array structure, (iii) Memory array structure with transistor at BL and (iv) Columns isolated array structure

Structure	Operation	V_{WL1}	V_{WL2}	V_{BL1}	V_{BL2}
Cross-point memory array [10, 24]	Write 1	Vdd	0.5 Vdd	GND	0.5 Vdd
	Write 0	GND	0.5 Vdd	Vdd	0.5 Vdd
	Read	±Vdd	Floating	To sense amp.	Floating
Memory array with transistor at BL [5, 15, 17]	Write 1	Vdd	Floating	GND	Floating
	Write 0	−Vdd	Floating	GND	Floating
	Read	±Vdd	GND	To load	To load
Columns isolated memory array [16, 18]	Write 1	V_{dd}	*Floating*	*GND*	*Floating*
	Write 0	$-V_{dd}$	*Floating*	*GND*	*Floating*
	Read	$\pm V_{dd}$	*Floating*	*GND*	*Floating*

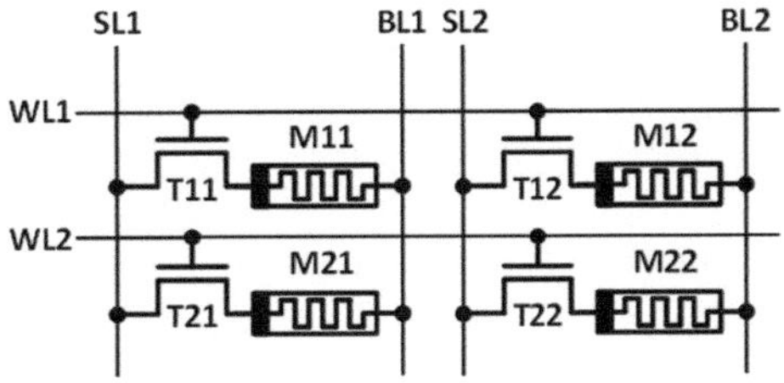

Fig. 2 Two-input LUT using MOS accessed memory array structure

To select a memristor (M11) for a WRITE or READ operation requires the transistor (T11) in the corresponding word line (WL1) to be switched ON. Then the WRITE or READ voltage is applied to SL1. To unselect the other memristors, the remaining word line (WL2), select line (SL2) and bit line (BL2) are connected to 0.5 Vdd. Both the WRITE and READ operations should ensure that the NSP of the unselected memristor cells do not change over a specified number of write cycles or a specified time with 0.5 Vdd applied. Otherwise applying refresh pulse is unavoidable. Also each cell structure is large due to the MOS device that is necessary to drive enough write current even though the memristor cell itself is much smaller.

(iii) *Memory array structure with transistors at BL* [5, 15, 17]

This architecture is similar to cross-point memory structure but uses only one transistor at for each BL. Therefore it reduces the area when compared with the structures presented in [10, 11]. In addition, a new WRTIE and READ method is proposed; for WRITE operation, applying +Vdd (+0.9 V) and –Vdd (–0.9 V) are applied only to WL. Therefore unlike previous NV LUT structures [10, 11] it does not require an additional circuit to monitor the incoming data and then channel it to WL/BL depending on the value of the data for the WRITE operation. In addition, this NV LUT structure does not require the application of +Vdd and +0.5 Vdd to

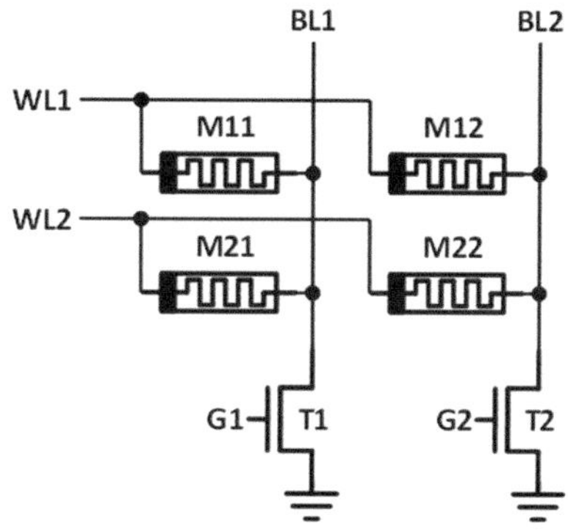

Fig. 3 Two-input LUT using memory array structure with transistors at BL

WL and BL to unselect a memristor, therefore, the presented NV LUT structure uses only three power rails (+Vdd, −Vdd and GND). Also the variations in NSP of the unselected memristors are small when compared with the previous structures [10, 11] (Fig. 3).

In this structure, a memristor (M11) is selected for WRITE or READ operation by applying the appropriate voltage to the WL(WL1) and turning on appropriate transistor (T1). The WRITE and READ voltages are presented in Table 1. For WRITE 1 (+Vdd) or 0 (−Vdd) operation appropriate voltage is applied to the corresponding WL until the NSP of the memristor is completely programmed. For READ 1(0) operation, a pulse of +Vdd (10 ns) is applied to WL (READ pulse) of the memristor with NSP at 1(0) and the READ value is found across the NMOS. As shown in the next section, the READ 1 operation did not affect the NSP of the memristor. Therefore, unlike previous structures [10, 11] this structure does not require a RESTORE pulse following the READ 1 operation. Thus, the presented NV LUT structure method accomplishes 25% READ and RESTORE time savings when compared with [10, 11]. However, during a READ 0 operation, the NSP of the memristor slightly increases towards Ron. Therefore a RESTORE pulse −Vdd is applied for 10 ns to restore the NSP to its original value, but the RESTORE pulse width is insufficient to restore the NSP to its original value. Therefore to study the effect of NSP with continuous READ 0 operations, consecutive READ 0 and RESTORE operations (~400 cycles) with an initial value of NSP equal to 0 has been performed. As illustrated in Fig. 4, the NSP does not reach the threshold values; instead an interesting ringing behaviour has been observed for the NSP with an average value well below the threshold value (0.5) when the time of the READ operation is equal to the time for the RESTORE operation. While a similar testing on the structure proposed in [10] causes the NSP to cross the threshold value as shown in Fig. 5. In addition, it can be noted as the READ pulse width decreases the ringing amplitude also decreases, so using smaller READ pulse width is better. More detailed analysis is presented in [5, 15, 17].

Also in this structure, for the READ operation a grounded row method is adopted that is able to provide a clear distinction between the memristors with NSP

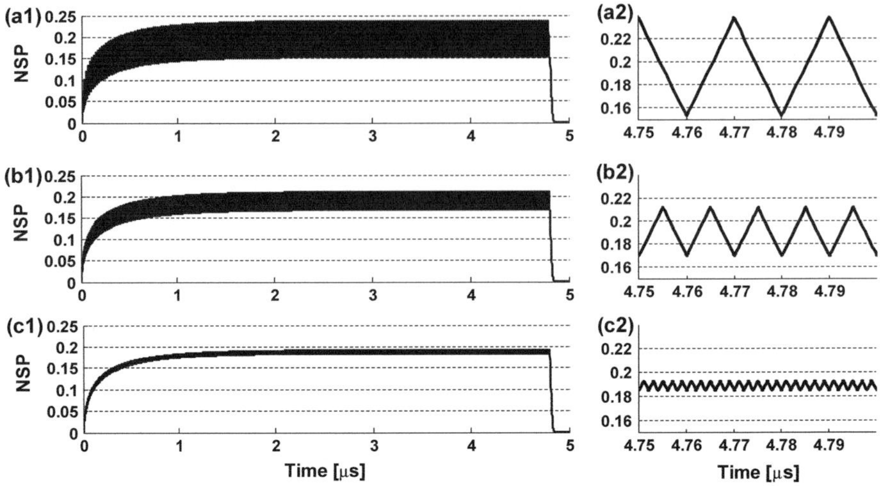

Fig. 4 NSP for consecutive READ 0 and RESTORE operations; **a1** READ pulse width of 10 ns, **a2** zoomed version of (**a1**), **b1** READ pulse width of 5 ns, **b2** zoomed version of (**b1**), **c1** READ pulse width of 1 ns and **c2** zoomed version of (**c1**)

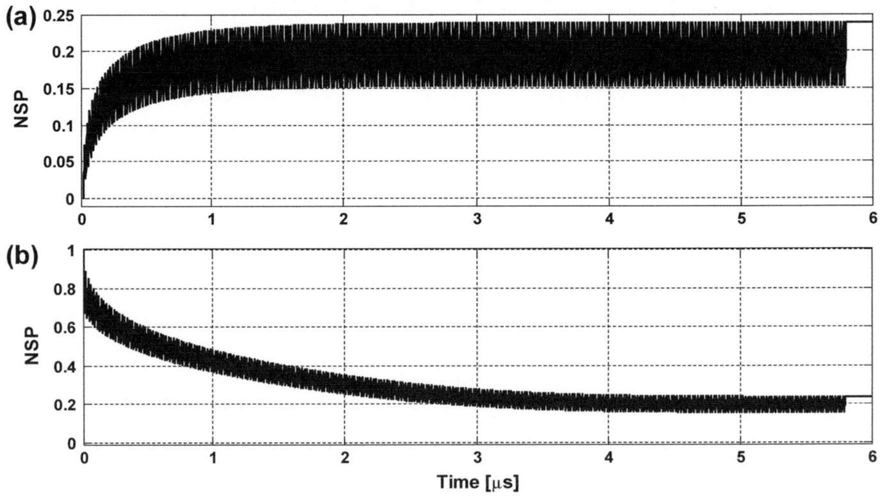

Fig. 5 For Structure [10], NSP for consecutive **a** READ 0 and RESTORE operations, **b** READ 1 and RESTORE operations

at 0 and 1 when compared with the floating row method that is being adopted by the structure [10].

Next is the NSP variation of the unselected memristors of this structure due to sneak path current is presented. Considering all memristors are at Roff state, the

WRITE 1 and 0 operations are sequentially performed on M11, this has resulted the NSPs of the unselected memristors M21 and M12 show a slight change towards Ron (well below threshold value) due to sneak path current generated during the WRITE 1 operation and no change during the WRITE 0 operation, respectively. Hence, the correct logic value (0) is stored in M21 and M12. Also the NSP of the unselected memristor M22 does not change during the WRITE 1 operation, but it changes significantly towards Ron during the sneak path current due to WRITE 0 operations. Thus the sneak path current affects the unselected memristors in this structure. Hence to overcome this, the structure executes in two phases WRITE operations. During the Phase 1 (Refresh Phase) all the memristors are written to 0 initially. During the Phase 2 all the selected memristors are subjected to WRITE 1 operation. Though this structure provides a better solution to handle the sneak path current effect on unselected memristors when compared with the structure presented in [10, 11] but at the cost of additional overhead, i.e. a refresh cycle for each WRITE operation.

(iv) *Columns isolated memory array structure* [16, 18]

In this structure, LUT design uses memristors but without employing a nano-crossbar. As shown later in this paper, when compared with other structures, it has a fast WRITE time, significantly reduced READ power dissipation and no power dissipation in the stand-by mode. More importantly, it eliminates the effects of sneak path current, write half-select, while preserving significant performance features. In addition, this structure has better controllability of the NSPs for the unselected memristors.

The two-input LUT is shown in Fig. 6. It consists of columns of nano wires (BL). A terminal of every memristor is connected to a BL, while the other terminals of the memristors are connected to the controller (controller details is presented in [18]), thus forming independent horizontal lines that are still referred as WL. Unlike other structures, the memristors on a row are not connected. Thus the presented NV LUT structure does not utilize a nano-crossbar structure. The number of memristors connected to each BL determines the dimension of the LUT. Every BL is connected to ground through a MOS transistor (T1 and T2). According to the

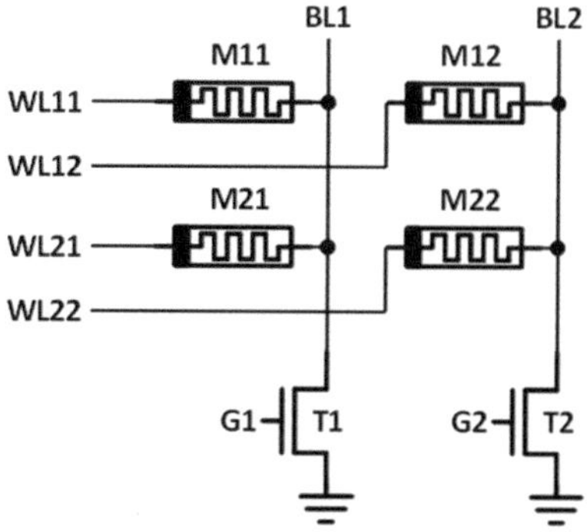

Fig. 6 Two-input LUT using column isolated memory array structure

input data, the controller handles the data to be driven on WL. Also, it switches on and off the transistors by controlling the gate signals (G1 and G2) and selects appropriate BL value to the output. The voltage requirements for the signals of the WRITE and READ operations of a selected memristor (M11) are shown in Table 1.

The WRITE operation is performed across all memristors connected to a particular BL. By driving T1(T2) to on state, the memristors connected to BL1 (BL2), i.e. M11 and M21(M12 and M22) are involved in the WRITE operation. So depending on the WRITE data (Table 1) the corresponding value is written in the memristors. As shown in Fig. 6, BL is connected to ground through the MOS transistors. Hence when performing the WRITE operation on the memristors connected to BL1, the memristors connected to BL2 are unaffected because those memristors are totally independent of BL1 and T2 is turned off. Therefore unlike previous schemes, the presented NV LUT structure does not suffer from the write half-select problem; in addition, the presented NV LUT structure does not require a 0.5 Vdd bias to unselect a memristor and a two-step writing scheme. The WRITE is performed across the memristors connected to a BL, so the WRITE delay is significantly reduced.

The READ operation is performed on the memristor (M11) by tuning on T1, while the remaining memristors on BL1 and BL2 are connected to a high resistance (floating) because the corresponding pass transistors (TG2, TG3 and TG4) (refer to the controller [18]) are in the off state. By applying the READ voltage (Table 1) on WL1 and propagating the voltage across T1 the value stored at M11is read out. Furthermore, the output voltage difference between the READ 0 (NSP = 0) and READ 1 (NSP = 1) in this structure is significantly greater than the other structures.

Next the effect of sneak current path on the unselected memristors is presented. As it is explained earlier, the BL are independent to each other, so the unselected memristors (R12 and R22) that are connected to the BL2 are not affected during the READ operation on R11. This is applicable to any LUT size. Consider next the unselected memristor (R21) that is connected to the same BL to which the selected memristor (R11) is connected. R21 is unselected by turning off the pass transistors (TG3) (refer to the controller in [16]) that provides a very high resistance to its path. Therefore the current through WL21 is infinitesimally small and hence the NSP of the unselected memristor is unaffected. Thus the presented NV LUT structure does not suffer from the sneak path current problem. As the memristors in the rows are not connected, 0.5 Vdd biasing is not required and therefore the presented NV LUT structure does not incur the write half-select problem.

The complexity of presented structures in terms of number of control lines, inputs and transistors are presented in Table 2. Among all the structures, the column isolated memory array has more number of inputs, but considering the maximum commercially available LUT inputs (six) and each LUT inputs are have short routing length to the controller, the column isolated memory array provides better performance.

Table 2 Number of control lines, inputs and transistors for an nxn LUT memory under various LUT structures

Structure	Number of control lines	Number of inputs	Number of required transistors
Cross-point memory array [10, 24]	$2n$	$2n$	$2n$
MOS accessed memory arrays [11]	n	$2n$	n^2
Memory array with transistor at BL [5, 15, 17]	n	n	n
Columns isolated memory array [16, 18]	n	n^2	n

4 Simulation Results

The SPICE simulation results of the LUT with dimension of 2, 4, 6 and 8 designed using Crossbar memory array [10, 24], Memory array with transistor at BL [5, 15, 17] and Columns isolated memory array [16, 18] are analyzed. The parameters used are given as follows: Memristor: Ron = 100 Ω; Roff = 19 kΩ, D = 10 nm; Transistors: Low Power (LP) model of the Predictive Technology Model (PTM), gate length = 32 nm; Width = 1 μm; gate voltage = 0.9 V; Accuracy of NSP: 0.1%; Simulator: LTSPICE IV. Different scenarios for the WRITE and READ operations [16] considered are.

- *Scenario 1*: *W1*-WRITE 1 to all memristor when the NSPs of all memristors are initially 0.
- *Scenario 2*: *W0*-WRITE 0 to all memristor when the NSPs of all memristors are initially 1.
- *Scenario 3*: *W0A1*-WRITE 0 to a memristor while the NSPs of all memristors are initially 1.
- *Scenario 4*: *W1A0*-WRITE 1 to a memristor while the NSPs of all memristors are 0.
- *Scenario 5*: *R0A1SC*-READ 0 when only the NSP of the selected memristor is 0 (i.e. the NSP of all other memristors are 1).
- *Scenario 6*: *R0A0SC*-READ 0 when the NSPs of all memristors are 0.
- *Scenario 7*: *R1A0SC*-READ 1 when only the NSP of the selected memristor is 1 (i.e. the NSP of all other memristors are 0).
- *Scenario 8*: *R1A1SC*-READ 1 when the NSPs of all memristors are 1.

(a) *WRITE operation*

The average and worst case write delay, energy dissipation and energy delay product (EDP) of the LUT for mentioned four structures obtained using the four WRITE operation scenarios are presented [18] in Figs. 7 and 8.

The average and worst case WRITE delays of Column isolated memory array-based LUT are significantly less than the other LUT structures. In addition, as the dimension of the LUT increases the difference between the WRITE delay for the

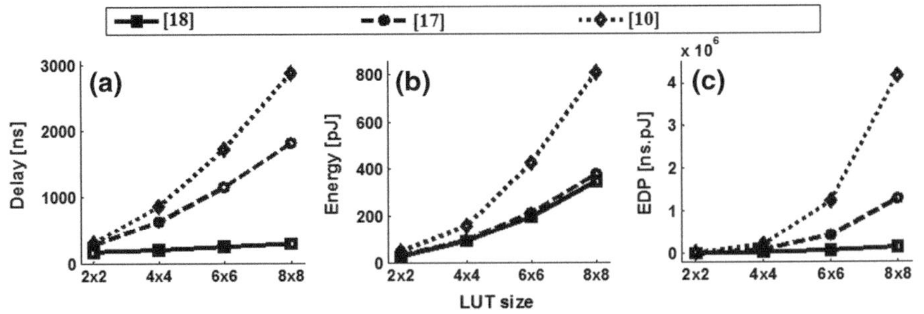

Fig. 7 Average WRITE performance versus LUT size; **a** delay, **b** energy and **c** EDP [18]

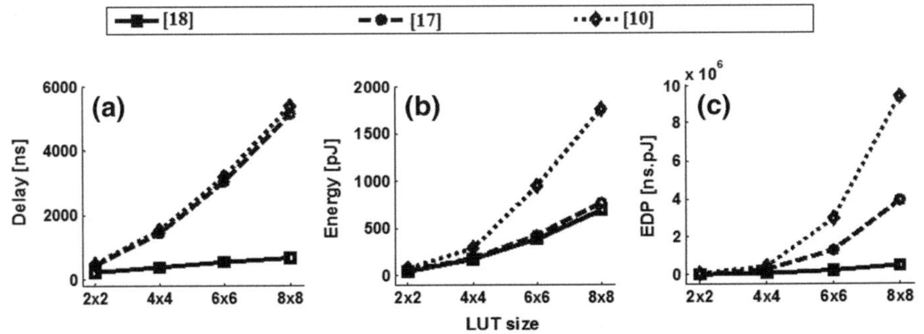

Fig. 8 Worst-Case WRITE performance versus LUT size; **a** delay, **b** energy and **c** EDP [18]

Column isolated memory array-based LUT and other LUT structures increases significantly. This occurs because, in the Column isolated memory array-based LUT, the WRITE operation is performed across all the memristors connected to a BL. Also, the average and worst case EDPs of the Column isolated memory array-based LUT structure is significantly less than other LUT structures.

(b) *READ operation*

The average and worst case READ delay, energy dissipation and energy delay product (EDP) of the LUT for mentioned four structures obtained using the four READ operation scenarios are presented [18] in Figs. 9 and 10.

The average and worst case READ delays remain constant, so nearly independent of LUT dimension for the Column isolated memory array-based LUT structure. This is caused by the constant value of the load resistance. In addition, when compared with other LUT structures, the average and worst case READ delays of the Column isolated memory array-based LUT structure are significantly decreased, so capable of delivering a significantly faster READ operation as very important feature in a FPGA. Also, the average and worst case EDPs of the Column isolated memory array-based LUT structure are significantly less than other LUT structures.

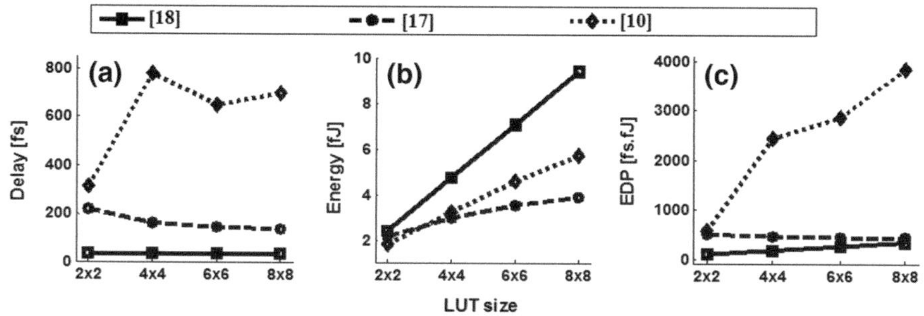

Fig. 9 Average READ performance versus LUT size; **a** delay, **b** energy and **c** EDP [18]

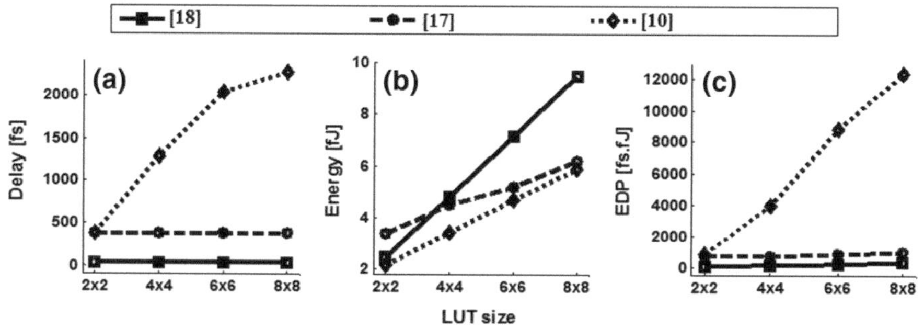

Fig. 10 Worst-Case READ performance versus LUT size; **a** delay, **b** energy and **c** EDP [18]

(c) *Sneak path effect on NSP variation* [16, 18]

The changes to the NSPs of the unselected memristors of a LUT designed using four structures for a WRITE operation are assessed by simulation at different sizes of the LUT. The simulation results are presented in Table 3. As expected, the NSPs of the unselected memristors in the Column isolated memory array-based LUT structure do not change during both the WRITE 1 and WRITE 0 operations; this condition has been verified for different sizes of LUTs (up to dimension 8). This mainly contributed because of the isolated BLs. The NSPs of other some of the unselected memristors in the other LUT structures [10, 17] change and the magnitude of this change is such that the threshold value is often reached. For WRITE 1 operation, the results of [10] show the loss of data integrity. Similarly for WRITE 0 operation [10, 17] losses the data integrity. Thus, the Column isolated memory array-based LUT structure in addition to preventing sneak path currents, also preserves data integrity.

Table 3 Comparison of NSP change during WRITE operation [18]

LUT size	WRITE 1 operation						WRITE 0 operation					
	No. of affected (unselected) memristors			Worst case NSP among unselected memristors with initial value of 0			No. of affected (unselected) memristors			Worst case NSP among unselected memristors with initial value of 1		
	[18]	[17]	[10]	[18]	[17]	[10]	[18]	[17]	[10]	[18]	[17]	[10]
2 × 2	0	2	1	0	0.179	0.868	0	2	2	1	0.175	0.016
4 × 4	0	6	3	0	0.242	0.838	0	6	6	1	0.173	0.048
6 × 6	0	10	5	0	0.261	0.827	0	10	10	1	0.127	0.093
8 × 8	0	14	7	0	0.269	0.812	0	14	14	1	0.173	0.140

5 Conclusion

This paper has presented detailed over view of different memristor-based nano-crossbar structures (Cross-point memory array structure, MOS accessed memory array structure, Memory array structure with transistors at BL and Columns isolated memory array structure.) that are employed for designing the NV LUT of a FPGA. Particular focus is on efficient READ and WRITE operations; techniques to overcome the sneak path current. Among the LUT structures, Columns isolated memory array structure, eliminates the effect of sneak path current on the unselected memristors. It has a fast WRITE time, significantly reduced READ power dissipation and no power dissipation in the stand-by mode. In addition, it eliminates the write half-select issue. Moreover, it prevents the data integrity when compared with other structures as this structure has better controllability of the NSPs for the unselected memristors.

Acknowledgements This work is based on the contributions from Prof. Fabrizio Lombardi, Northeastern University, Boston USA and Prof. Haider Abbas, The University of Nottingham, Malaysia.

References

1. H.A.F. Almurib, T.N. Kumar, F. Lombardi, A single-configuration method for application-dependent testing of SRAM-based FPGA interconnects, in *Proceedings of the 20th IEEE Asian Test Symposium, New Delhi* (2011), pp. 444–450
2. T. Nandha Kumar, F. Lombardi A novel heuristic method for application dependent testing of a FPGA interconnect. IEEE Trans. Comput. **16**(1), 163–172 (2013)
3. K.J. Han et al., A novel flash-based FPGA technology with deep trench isolation, in *22nd IEEE Non-Volatile Semiconductor Memory Workshop* (2007), pp. 32–33
4. Xilinx SpartanTM-3AN FPGAs, http://www.xilinx.com
5. T.N. Kumar, H.A.F. Almurib, F. Lombardi, On the operational features and performance of a memristor-based cell for a LUT of an FPGA, in *Proceedings of the13th IEEE International Conference on Nanotechnology* (2013), pp. 71–76

6. L.O. Chua, Memristor—the missing circuit element. IEEE Trans. Circ. Theory **ct-18**(5), 507–519 (1971)
7. D.B. Strukov, G.S. Snider, D.R. Stewart, R.S. Williams, The missing memristor found. Nature **453**, 80–83 (2008)
8. J. Cong, B. Xiao, mrFPGA: a novel FPGA scheme with memristor-based reconfiguration, in *Proceedings of the IEEE/ACM International Symposium on Nanoscale Architectures* (2011), pp. 1–8
9. S. Tanachutiwat, M. Liu, W. Wang, FPGA Based on Integration of CMOS and RRAM. IEEE Trans. Very Large Scale Integr. VLSI Syst. **19**(11), 2023–2032 (2011)
10. Y. Ho, G.M. Huang, P. Li, Dynamical properties and design analysis for nonvolatile memristor memories, in IEEE Trans. Circ. Syst. I **58**(4) (2011)
11. C. Xu, X. Dong, N.P. Jouppi, Y. Xie, Design implications of memristor-based RRAM cross-point structures, in *Proceedings of the Design, Automation and Test in Europe* (2011), pp. 1–6
12. A. Chen, Accessibility of nano-crossbar arrays of resistive switching devices, in *Proceedings of the IEEE International Conference on Nanotechnology* (2011), pp. 1767–1771
13. I.E. Ebong, P. Mazumder, Self-controlled writing and erasing in a memristor crossbar memory, in *IEEE Transactions on Nanotechnology*, vol. 10, no. 6 (2011)
14. J. Liang, H.-S. P. Wong, Cross-point memory array without cell selectors—device characteristics and data storage pattern dependencies. IEEE Trans. Electron Dev. **57**(10) (2010)
15. H.A.F. Almurib, T.N, Kumar, F. Lombardi, A memristor-based LUT for FPGAs, in *Proceedings of the 9th IEEE International Conference on Nano/Micro Engineered and Molecular System IEEE-NEMS* (2014), pp. 448–453
16. T.N, Kumar, H.A.F. Almurib, F. Lombardi, A novel design of a memristor-based look-up table (LUT) for FPGA, in *IEEE Asia Pacific Conference on Circuits and Systems* (2014)
17. H.A.F. Almurib, T. Nandha Kumar, F. Lombardi, Design and evaluation of a memristor-based look-up table for non-volatile field programmable gate arrays. IET Circ. Dev. Syst. (2016)
18. T. Nandha Kumar, H.A.F. Almurib, F. Lombardi, Design of a memristor-based look-up table (LUT) for low-energy operation of FPGAs. Integr. VLSI J. (2016). Elsevier
19. P.W.C. Ho, T. Nandha Kumar, H. Abbas, Configurable memristive logic block for memristive-based FPGA architectures. Integr. VLSI J. (2016). Elsevier
20. P.W.C. Ho, H.A. Mohamed, T. Nandha Kumar, One-bit non-volatile memory cell using memristor and transmission gates, in *IEEE International Conference on Electronic Design (ICED 2014), Malaysia* (2014), pp. 244–248
21. P. Junsangsri, F. Lombardi, A novel hybrid design of a memory cell using a memristor and ambipolar transistors, in *Proceedings of the 11th IEEE International Conference on Nanotechnology* (2011)
22. A. Chen, Accessibility of nano-crossbar arrays of resistive switching devices, in *IEEE International Conference on Nanotechnology* (2011)
23. M.M. Ziegler, M.R. Stan, Design and analysis of crossbar circuits for molecular nanoelctronics, in *IEEE Nano* (2002)
24. N.Z. Haron, S. Hamdioui, On defect oriented testing for hybrid CMOS/memristor memory. Proc. ATS (2011)
25. Y. Chen, J. Zhao, Y. Xie, 3D-NonFAR: three-dimensional non-volatile FPGA architecture using phase change memory, in *2010 ACM/IEEE International Symposium on Low-Power Electronics and Design (ISLPED)* (2010), pp. 55–60
26. S. Tanachutiwat, M. Liu, W. Wang, FPGA based on integration of CMOS and RRAM. IEEE Trans. Very Large Scale Integr. (VLSI) Syst **19**(11), 2023–2032 (2011)
27. Y.Y. Liauw, Z. Zhang, W. Kim, A.E. Gamal, S.S. Wong, Nonvolatile 3D-FPGA with monolithically stacked RRAM-based configuration memory", in *IEEE International Solid-State Circuits Conference Digest of Technical Papers (ISSCC)* (2012), pp. 406–408

28. M. Lin, A. El Gamal, Y.-C. Lu, S. Wong, Performance benefits of monolithically stacked 3-D FPGA. IEEE Trans. Comput. Aided Des. Integr. Circ. Syst. **26**(2), 216–229 (2007)
29. Y. Meng, T. Sherwood, R. Kastner, Leakage power reduction of embedded memories on FPGAs through location assignment, in *43rd ACM/IEEE Design Automation Conference* (2006), pp. 612–61
30. D. Lewis et al., Architectural enhancements in stratix-III TM and stratix-IV TM, in *Proceedings of the ACM/SIGDA International Symposium on Field Programmable Gate Arrays (FPGA '09)* (ACM, New York, 2009), pp. 33–42
31. O. Turkyilmaz, S. Onkaraiah, M. Reyboz, F. Clermidy, Hraziia, C. Anghel, J.M. Portal, M. Bocquet, RRAM-based FPGA for "Normally Off, Instantly On" applications, in *IEEE/ACM International Symposium on Nanoscale Architectures* (2012), pp. 101–108

Chapter 9
Radio-Frequency Integrated Circuit Design

Jie Li

Abstract The boom of wireless and mobile networks has led to an ever-increasing demand for high performance, low-power, and low-cost radio-frequency integrated circuit (RFIC) design. Advances in silicon and silicon–germanium-based technologies can now provide highly integrated system-on-chip (SOC). In spite of this significant motivation, this talk provides information about the fundamentals of RFIC design for wireless communication systems. The talk starts with a discussion on the potential applications, the technology trade-offs, and challenges on the RFIC design. We then focus on wireless transceiver IC designs with the brief analysis for each analogy front-end block, such as low-noise-amplifier (LNA), mixer, and voltage-controlled oscillator (VCO) designs. We will also present some essential topics in RFIC testing and practical implementations.

Keywords Radio-frequency integrated circuit (RFIC) · CMOS · Low-noise amplifier (LNA) · Mixer · Voltage-controlled oscillator (VCO)

1 Introduction

The future Internet, now is known as the "Internet of Things (IoT)" is foreseen to be a world-wide network of interconnected objects uniquely addressable, based on standard communication protocols. Identified by a unique address, any object including computers, sensors, radio-frequency identification (RFID) tags, or mobile phones will be able to dynamically join the network, collaborate, and cooperate efficiently to carry out different tasks. Billions of devices are predicted to be connected to the world in the coming years. This has also raised concern about the sustainability of the growing energy footprint of communication networks.

J. Li (✉)
Electrical and Electronic Engineering Department, University of Melbourne, Melbourne, Australia
e-mail: Jie.li@unimelb.edu.au

J. Li
CTO, Graham Innovations Ltd., Melbourne, Australia

S.R.S. Prabaharan et al. (eds.), *Frontiers in Electronic Technologies*,
Lecture Notes in Electrical Engineering 433, DOI 10.1007/978-981-10-4235-5_9

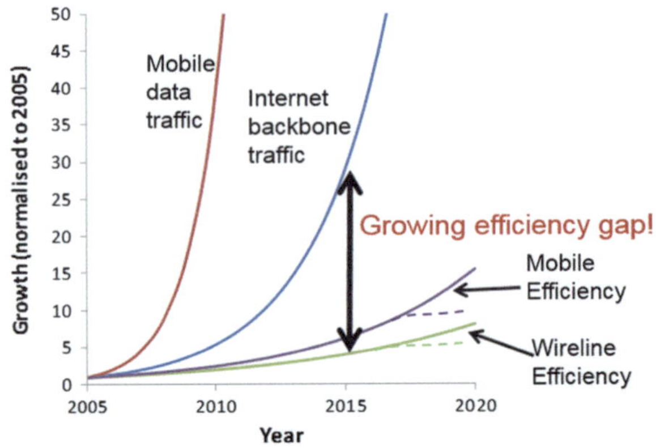

Fig. 1 Growing efficiency gap in the current networks [1]

Figure 1 shows the challenge we face in our endeavors to make the Internet and ICT sustainable [1]. This plot shows the annual increase in mobile traffic and Internet backbone are 108% and 34% per annual, respectively. The trends for improving energy efficiency are only 15% and 20% for the wireline and mobile network, respectively. We see that traffic growth is far out-stripping efficiency improvements. The difference is called the "efficiency gap". Unless this gap can be closed, long term sustainability of the Internet and ICT are not possible for the arrival of the IoT era. Researches and scientists has developed and integrated green technologies from both device-level and circuit-level to counter this challenge [2, 3].

As one of the main members in the IoT ecosystem, the RFID device serves the same purpose as a bar code or a magnetic strip on the back of a credit card or product label, proving a unique identifier for that object. Considering the quantity and cost of the RFID devices, we need to optimize its power consumption, simplify its architecture and minimize its chip size while maintain superior performance meeting all design specifications. Applying new technologies on its radio-frequency (RF) transceiver is one of the most fundamental approaches.

2 An Example of RF Architecture for the RFID Receiver

Among all the RF receiver architecture topologies, the direct conversion receiver, also known as the homodyne or zero-IF receiver, are well used in low-power applications as only a single mixing stage is required for the signal conversion. As mixers often occupy the largest proportion of the receiver front-end's power, less mixing stage reveals less power consumption. The direct conversion architecture topology is shown in Fig. 2 in a typical RF application. To overcome the mixing sideband image issue, the in-phase/quadrature (I/O) conversion with frequency-modulation structure is employed in many recent RF receiver. In this

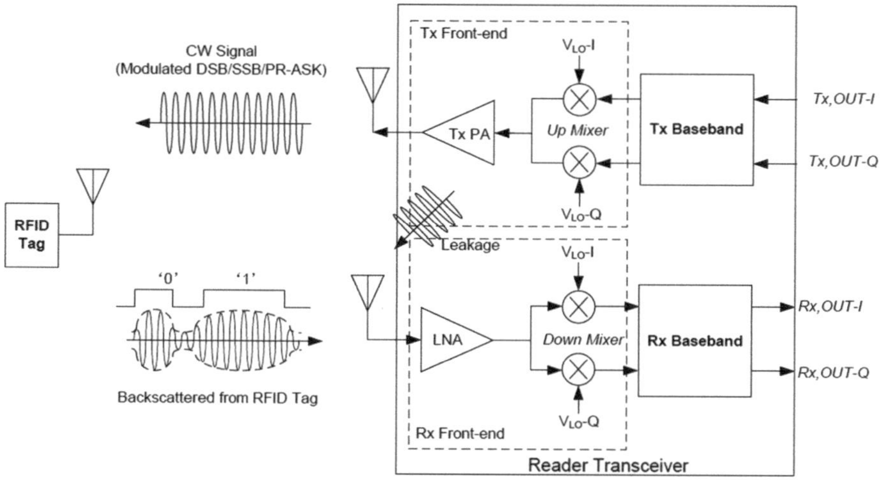

Fig. 2 The front-end block of a typical RFID reader based on the single-conversion receiver with frequency modulation (FM) and phase modulation (PM)

talk, we use this topology as an example for a RFID receiver design to provide a good trade-off between power consumption, linearity, noise, and sensitivity.

As the size of transistor has moved into the deep submicron region, it is not only difficult to maintain the fabrication accuracy, but also brings the challenge for RF design engineers to simulate and predicate the performance of the device accurately. The use of complicated design models brings in significant timing and development cost. The model with the inclusion of the finite drain-source conductance (g_{ds}) effect can significantly improve the accuracy of simulation while maintaining a realistic simple model for nanometric design [4]. This is because the finite output conductance is no longer negligible compared to the transconductance (g_m) for low-power deep nanometric CMOS devices. We include the g_{ds} in the design consideration for the proposed RF analog front-ends, including the low-noise amplifier (LNA), the quadrature mixer, and the voltage-controlled oscillator (VCO).

3 An Example of LNA Design for the RFID Receiver

Being the first block in the receiver front-end chain, the LNA determines the noise figure (NF) of the receiver, in other words, revealing the sensitivity and bit-error rate of the system. Various kinds of LNAs are available for different applications, such as narrow band, multiple bands, and wide-band LNAs. RFID applications are typically based on a specified frequency, therefore, a narrow-band LNA tuned at a certain frequency is used in our RFID reader design example. Several optimization techniques have been developed for narrow-band LNA with respect to power dissipation, noise, and impedance matching [5, 6]. A proposed technique, namely enhanced power-constrained simultaneous noise and input matching technique (PCSNIM), is used as an example to demonstrate how impedance and minimum noise matching can be achieved simultaneously at the very low power consumption and power supplier [4].

Figure 3 shows the circuit diagram of the proposed ultra-high frequency (UHF) RFID LNA based on the enhanced PCSNIM technique. M1 with L_s is the input inductive source degenerated amplifier device which creates a noise-free resistive component considering the gate input facilitating matching with the source resistance. L_g is an external inductor for complex conjugate input reactance matching. C_e is an external capacitor to facilitate matching for low-power design. For low-power deep nanometric CMOS devices, the finite output conductance ($g_{ds} = 1/r_o$) is shown in parallel with the MOS devices. L_d, R_d, and C_d form the output tank circuit tuned at the operating frequency and the relative values of L_d and C_d are carefully chosen to prevent tank detuning by the parasitic capacitance at the drain of M2. C_{ac} at the input and at the output are AC-coupling capacitors, while C_b is an AC-grounding capacitor at the gate of the cascoding device M2. M3 in conjunction with R_{ref} controls the bias current through the telescopic cascode while a large R_b ensures that noise and any AC signal pick-up at the gate of the current mirror device M3 is largely decoupled from the RF input at the gate of M1.

Table 1 shows a summary of performance comparison of the proposed UHF LNA with other UHF LNA designs indicating a very low NF achieved at sub-mW power. We used the follow figure of merit (FOM) for the performance

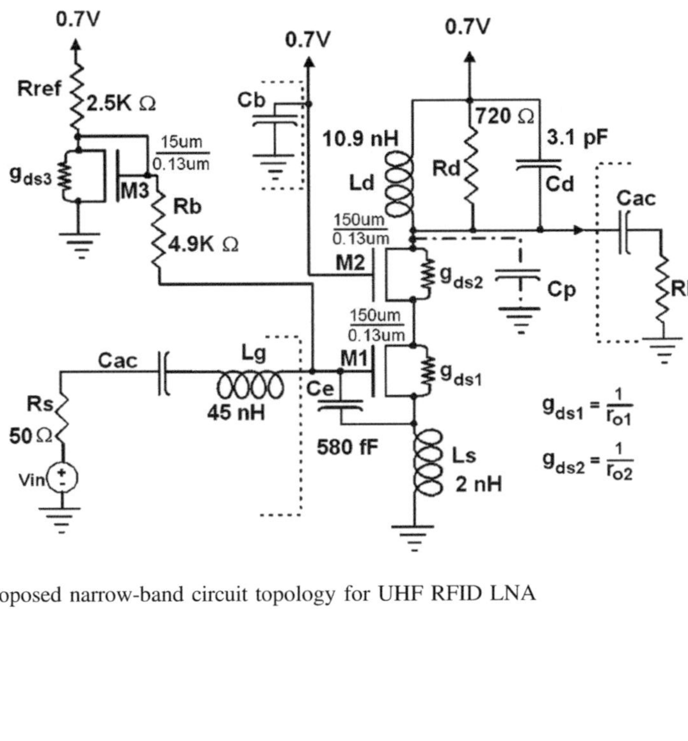

Fig. 3 Proposed narrow-band circuit topology for UHF RFID LNA

Table 1 Summary of the proposed UHF RFID LNA performance, and comparison with others

	CMOS process (nm)	Input isolation S_{11} (dB)	Voltage gain S_{21} (dB)	NF (dB)	Linearity IIP3/P1 dB (dBm)	P_{DC} (mW)	Supply voltage (V)	FOM
This work [4]	130	−30	17	2.2	−11.5/ −16.1	0.85	0.7	12.66
[7]	250	−18	12	1.35	−4/−15	2	1.25	5.45
[8]	350	−11	13.4	3.2	10.8/1/4	33	3.0	1.299
[9]	180	N/A	15	2.9	N/A/−15	4.32	1.8	1.37

comparison in Table 1 [5]. P_{DC} is the DC power dissipation of the LNA. The FOM clearly indicates the advantage of the presented RFID LNA design compared to the other designs.

$$\mathrm{FOM}[\mathrm{mW}^{-1}] = \frac{|\mathrm{Gain}|}{|NF - 1| \cdot P_{DC}[mW]}$$

4 An Example of Mixer Design for the RFID Receiver

Mixers translate base-band information between radiofrequency (RF) and intermediate frequency (IF) spectrums, and are integral part of any RF front-end. Direct conversion receiver (DCR) architecture is in favor by the IoT application design due to its high integration, low-cost and simplicity of base-band architecture. A double-balanced mixer is specifically suited to DCR due to low RF and local oscillator feed-through as well as high LO-to-IF isolation. Various techniques have been explored to optimize the overall performance among power consumption, flicker noise, thermal noise, conversion gain, and linearity. Current bleeding technique reduces the flicker noise contribution of the switching stage since the DC current through the switching devices is minimized [10]. However, the drawback of the bleeding current source is an increment in the tail capacitance, which reduces the common-mode rejection (contributes to nonlinearity) in addition to indirectly augmenting flicker noise. On the other hand, most of the RF signal current in this topology is forced into the low impedance source branches of the switching pairs since the current source (PMOS device) drain terminal has large output impedance. However, due to the finite g_{ds} effect, the signal leakage into the current bleeder is higher in nanometric CMOS processes and requires improvement to enhance the conversion gain. In terms of the thermal noise budget, the transconductor accounts for a significant proportion of it. This noise can be reduced significantly using an inductively degenerated merged LNA as the transconductor device.

An improved mixer using a novel technique of inductive degenerated current bleeding devices, and resonant tail capacitance tuning along with a matched input merged LNA/transconductor stage is proposed in this talk for the RFID reader receiver. Common-mode inter-modulation feedback is also implemented to improve linearity against DC level (device overdrive) perturbations.

Figure 4 shows the circuit topology of the proposed double-balanced LNA-merged mixer. M1 and M2 constitute the merged LNA/RF transconductors that are inductively degenerated by L_S (with half of L_S in each half circuit of the differential LNA) and simultaneously power and noise matched at UHF frequency through enhanced PCSNIM technique. L_{shunt} is an inductor for resonating the tail capacitance of the balanced switching pairs (with half of L_{shunt} in each half of the double-balanced mixer) to reduce flicker noise through indirect mechanism. L_S is ground center-tapped while L_{shunt} is AC-grounding center-tapped using C_{shunt}. The devices M3, M5, M4 and M6 constitute the double-balanced switching pairs commutating a small component of the DC current from the RF LNA/transconductor, while M7 and M8 are dual balanced PMOS current bleeding devices which channel the significant component (around 75%) of the DC bias drainage. However, with reduced bias current through the switching devices their drain impedances ($r_o = V_A/I_D$) are larger resulting in undesired RF leakage current through the bleeding devices, as well as, through the tail capacitance. In order to force most of the transconductor/LNA RF signal current (bias current perturbations) into the switching pairs, the impedance looking into the drain of the PMOS DC bleeder devices is boosted by inductive degeneration of the devices using L_{choke} (with half of L_{choke} with each bleeder device). L_{choke} also engages in parasitic tuning with the PMOS bleeding devices to ensure a very high impedance looking into the drain of the PMOS bleeders despite the finite g_{ds} of these nanometric devices. L_{choke} is a supply (V_{DD}) center-tapped inductor providing the high impedance RF blockage for the PMOS bleeders. Further inductive tuning to improve RF leakage and flicker noise performance can be enabled through bond-wire inductance (as well as an external inductor) in the vicinity of L_{choke} and L_{shunt} V_{DD} and AC-ground center-tap nodes, respectively. M9–M16 constitute the common-mode inter-modulation feedback to reduce DC bias (and overdrive) variation and improve the overall linearity of the mixer. For example, if the IF output common-mode levels go up, the current through diode-connected M9 goes up, resulting in the voltage across M9 to go up. Consequently, the bleeding current reduces resulting in the return to the preset common-mode output level through lowering of the DC levels. The mixer output load is provided by R_L instead of current source load to provide more signal headroom as well as minimize flicker noise.

Table 2 summarizes the proposed mixer's overall performance, in relation to some other recently published mixers, and, displays reported data from the cited designs. It is to be noted that some of the designs in the table are quadrature mixers and hence have some inherent performance differences with this double-balanced LNA-merged mixer. The proposed LNA-merged mixer provides an overall optimal performance based on power dissipation, supply voltage, NF, linearity, dynamic range and conversion gain suited to RFID application at UHF.

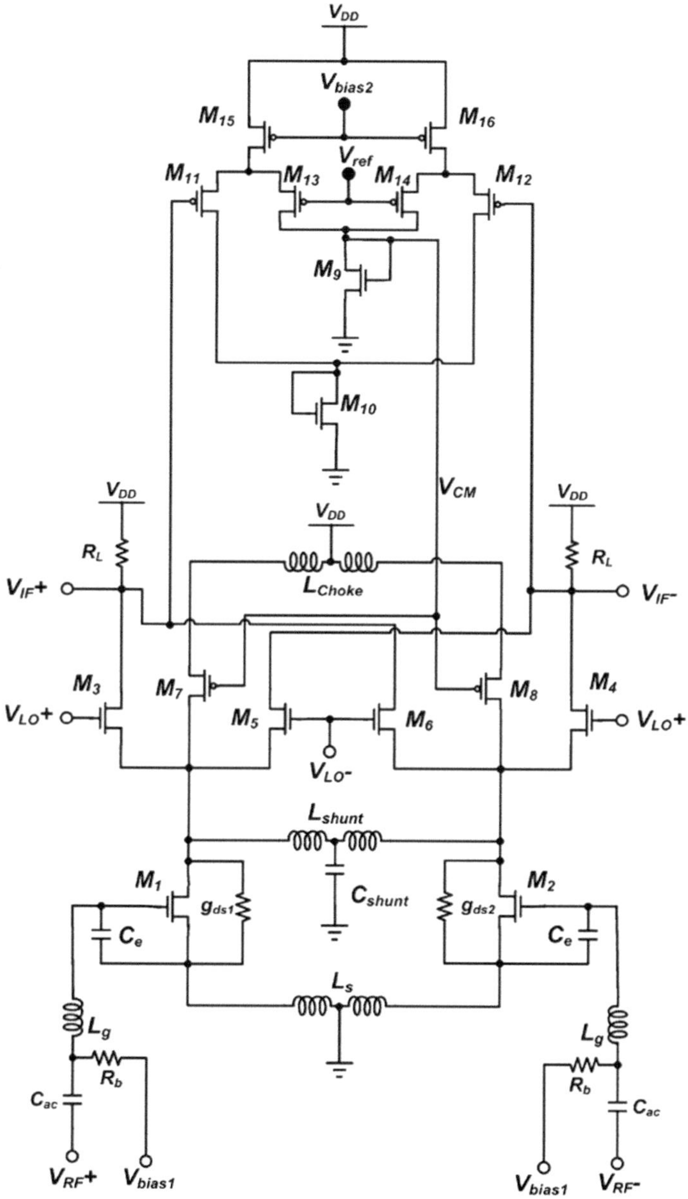

Fig. 4 Proposed merged LNA double-balanced mixer with inductive degenerated current bleeder devices, tail capacitance resonator, and common-mode inter-modulation feedback

Table 2 Summary of the proposed UHF RFID LNA-merged mixer performance in relation to other mixer designs

	CMOS process (nm)	Input isolation S_{11} (dB)	Conversion gain (dB)	NF_{NSB} (dB) @1 MHz	P1 dB (dBm)	IIP3/IIP2 (dBm)	P_{DC} (mW)	Supply voltage (V)
This work [11]	130	−19.5	17	7.5	−11.2	−4.1/37.4	6	1.2
[10]	130	−8.8	17.5	3.9	−10.5	0.84/N/A	34.5	1.5
[12]	130	N/A	7	6.5	N/A	2/60	2.1	2.7
[13]	90	<−10	12.1	8.4	−13	N/A/N/A	9.8	1.2
[14]	130	−22	15	15	N/A	−3/39	12	1.5

5 An Example of VCO Design for the RFID Reverie

VCOs in any RF front-end also consume a significant fraction of its power budget. Consequently, balancing the tradeoffs between power consumption, tuning range, tuning gain and phase noise is crucial in any VCO design. In addition, design of a quadrature VCO (QVCO) for image rejection in modulation and demodulation requires the consideration of quadrature error and possible tradeoff with phase noise and other performance parameters. Quadrature inaccuracies can result in residual image in a down-converted baseband mixer output. Various techniques are used for QVCO design, and the most well-known methods utilizes the cross and direct coupling of two LC-VCOs [15]. Coupling and biasing techniques dominate the trend in LC-QVCO in achieving low phase noise and low quadrature error.

Figure 5 shows the proposed LC-QVCO architecture with improved coupling and biasing techniques for two quadrature-coupled LC-VCOs in our RFID design [16]. The ports I+, I− and Q+, Q− are, respectively, the I-VCO and Q-VCO quadrature outputs along with their complementary signals. To improve the spectral purity of QVCO signals, transformer coupling has been introduced but with a trade-off for a large silicon area for UHF applications. Hence, in this design example, a pMOS folded-cascode structure is used to provide the mutual input excitations for quadrature generation. Compared with parallel quadrature coupling, series coupling using a cascode topology reduces the QVCO phase noise. In addition, the phase-noise performance is enhanced due to the lower hot carrier-induced drain current noise of the pMOS device in the folded cascode. A grounded center-tapped inductor is used for the LC-tank for power noise reduction power and linearity improvement in a mixed signal application. In addition, being ground center tapped, there is reduced substrate leakage currents from the inductor coils. This also results in the QVCO being less prone to self-resonance compared with other topologies with supply center-tapped or floating coils. The tail current source of the VCO-core positive feedback nMOS latch is eliminated to provide additional voltage headroom. Hence, it enjoys the advantage of oscillation under highly scaled supply voltage. The pMOS current sources biasing the cascode coupling topology also accounts for lower drain current noise compared with nMOS bias current sources due to its inherently lower noise behavior. To further reduce the noise of the pMOS current sources, pulsed biasing through complementary cross coupling is used in this design to release trapped carriers at the nSi-SiO interface. Pulsing the MOSFET overdrive between accumulation and strong inversion would thus free the carriers trapped during strong inversion resulting in reduced noise. Depletion-mode pMOS (D-PMOS) devices are used as varactor due to the quasi-linear monotonic capacitance versus voltage (C–V) behavior in this region of operation. Several devices are shunt-connected to increase the composite C–V tuning range. In addition, small-value metal–insulator–metal (MIM) capacitance is augmented in parallel with the D-PMOS array to increase the overall varactor linearity. Figure 1 also shows the DC connection polarity of the D-PMOS varactor with gates connected to the dc ground, while, the

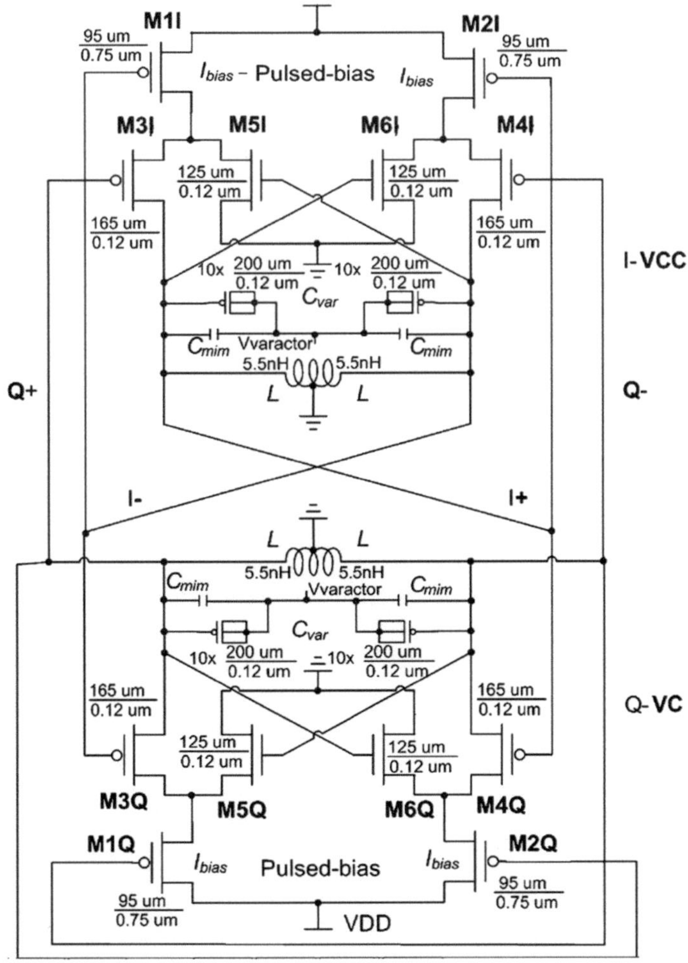

Fig. 5 Proposed low-voltage pulse-biased folded-cascode QVCO

shorted source, drain and body is connected to a variable positive voltage. In this topology, M5 and M6 constitute the core feedback latch devices of the VCOs. On the other hand, M3 and M4 are the corresponding coupling pMOS devices. The current flowing into the LC-tank is only that due to the drain current of the pMOS coupling devices. Since only one set of bias transistors consume drain current from the supply voltage, the operating supply voltage and power consumption can be significantly reduced compared to the conventional cross-coupled regeneration circuit.

Tuning changes the overall equivalent capacitance (C_{ov_eq}) consisting of the tank capacitance (C_{var}) and the capacitance looking between the ground and the drain of the coupling pMOS device. With large-signal operation, owing to the tail-free

Table 3 Summary of the proposed UHF RFID QVCO performance in relation to other mixer designs

	Process technology (nm)	Tuning range (%)	Phase noise (dBc/Hz) (MHz)	DC power (mW)	Supply voltage (V)
This work [16]	130	8%	−130@1	2.5	1
[17]	180	4.6%	−111@1	2.2	1
[18]	180	N/A	−124@1	8.7	1.45
[19]	130	N/A	−118@1	3.7	1.2
[20]	130	N/A	−112@1	4.8	1.2

grounded nMOS latch circuit, the pMOS cascoding devices and the cross-coupled nMOS devices are pulsed alternatively into accumulation (turned-OFF) and strong inversion (triode regime-ON), thus reducing the flicker noise contribution of these devices as well. The finite g_{ds} of the pMOS bias and the nMOS latch devices also act as transconductance degeneration for the pMOS coupling devices, resulting in a reduction of their transconductance, and, consequently, further reduction of their white noise contributions to the phase noise. Also, in this structure, the coupling and the regenerative devices are essentially in parallel for dc operation, but form a cascode structure for ac (large-signal nonlinear) operation, thus enabling the structure to have the benefits of both low phase noise and low quadrature error [10]. Furthermore, since the sustained oscillations only take place at a frequency for which the average magnitude of the negative resistance equals the parallel resistance of the LC-tank, no bimodal oscillation will occur in this circuit.

Table 3 shows a summary of performance comparison of the proposed QVCO with some other QVCO designs. With only 2.5 mW power consumption and −130 dBc/Hz SSB phase noise at 1 MHz offset from the carrier, this performance is quite competitive when compared to the recently published designs.

6 Summary of the RFID Receiver Design

The the completed RF receiver front-end for an IoT RFID application is presented based on the IBM 130 nm CMOS process. Figure 6 shows the complete layout diagram of the proposed RF front-end that includes single-end LNA, differential LNA, LNA-merged mixer, QVCO, and differential balun. The die diagram of each component is shown in Fig. 7. The whole receiver consumes less than 9 mW of power at 1.2 V supply voltage.

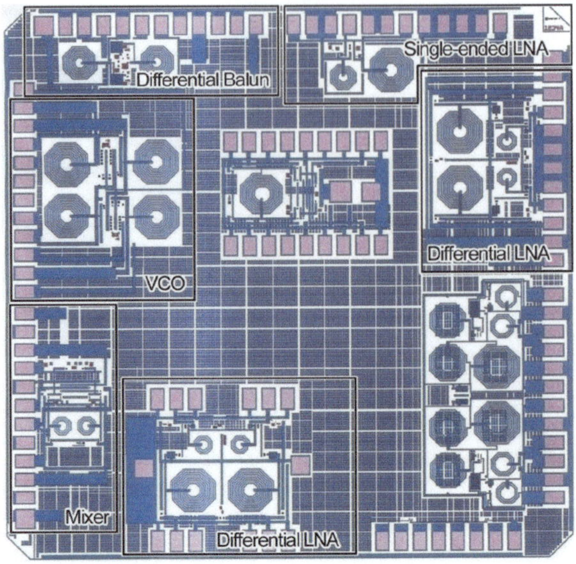

Fig. 6 Completed layout diagram of the RFID reader receiver front-end

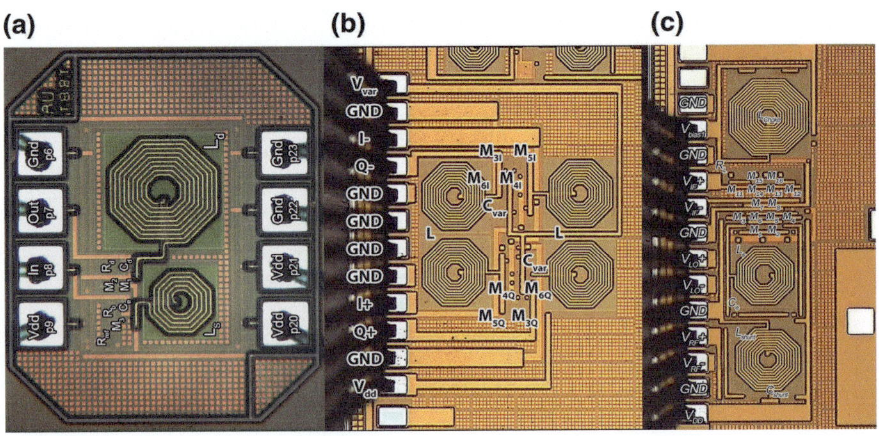

Fig. 7 Die photomicrograph of **a** single-ended LNA, **b** differential LNA-merged mixer, and **c** quadrature VCO of the fabriciated RFID receiver

References

1. K. Hinton et al., Power consumption and energy efficiency in the internet, in *IEEE Network*, vol. 25, no. 2, pp. 6–12, Mar-Apr 2011
2. S. Lambert et al., Road to energy-efficient optical access: greentouch final results. IEEE/OSA J. Opt. Commun. Netw. **8**(11), 878–892 (2016)

3. D. Kilper, Tutorial: energy efficient networks, in *2011 Optical Fiber Communication Conference and Exposition and the National Fiber Optic Engineers Conference, Los Angeles, CA*, 2011, pp. 1–67
4. J. Li, S.M.R. Hasan, Design and performance analysis of an 866-MHz low-power optimized CMOS LNA for UHF RFID. IEEE Trans. Ind. Electron. **60**(5), 1840–1849 (2013)
5. N. Trung-Kien et al., CMOS low-noise amplifier design optimization techniques. IEEE Trans. Microw. Theory Tech. **52**, 1433–1442 (2004)
6. M. Khurram, S.M.R. Hasan, A 3–5 GHz current-reuse g_{m}-Boosted CG LNA for ultrawideband in 130 nm CMOS, in *IEEE Transactions on Very Large Scale Integration (VLSI) Systems*, vol. 20, no. 3, pp. 400–409, Mar 2012
7. P.B. Khannur et al., A universal UHF RFID reader IC in 0.18 μm CMOS technology. IEEE J. Solid-State Circuits **43**(5), 1146–1154 (2008)
8. H.H. Roh et al., A common-gate low noise amplifier with high linearity over UHF RFID bands, in *Proceedings of Asia-Pacific* Symposium *on Electromagnetic Compatibility*, May 2008, pp. 88–91
9. S.P. Voinigescu et al., A scalable high frequency noise model for bipolar transistors with application optimal transistor sizing for low-noise amplifier design. IEEE J. Solid-State Circuits **32**(9), 1430–1439 (1997)
10. S.S.K. Ho, C.E. Saavedra, A CMOS broadband low-noise mixer with noise cancellation. IEEE Trans. Microw. Theory Tech. **58**(5), 1126–1132 (2010)
11. J. Li, S.M.R. Hasan, An inductive-degenerated current-bleeding LNA-merged CMOS mixer for 866 MHz RFID reader. Analog Integr. Circuits Signal Process. **80**(2), 173–185 (2014)
12. R.S. Pullela et al., Low flicker-noise quadrature mixer topology, in *IEEE International Solid-State Circuits Conference, ISSCC*, pp. 1870–1879, 2006
13. A. Amer et al., A 90-nm wideband merged CMOS LNA and mixer exploiting noise cancellation. IEEE J Solid State Circuits **42**(2), 323–328 (2007)
14. F. Yiping et al., Design of a high performance 2-GHz direct-conversion front-end with a single-ended RF input in 0.13 μm CMOS. IEEE J Solid State Circuits **44**(5), 1380–1390 (2009)
15. A. Rofougaran et al., A single-chip 900 MHz spread-spectrum wireless transceiver in 1 μm CMOS. I. Architecture and transmitter design. IEEE J. Solid-State Circuits **33**(4), 515–534 (1998)
16. J. Li, S.M.R. Hasan, Pulse-biased low-power low-phase-noise UHF LC-QVCO for 866 MHz RFID front-end. IEEE Trans. Microw. Theory Tech. **60**(10), 3120–3125 (2012)
17. T.-H. Huang, Y.-R. Tseng, A 1 V 2.2 mW 7 GHz CMOS quadrature VCO using current-reuse and cross-coupled transformer-feedback technology. IEEE Microw. Wirel. Compon. Lett. **18**(10), 698–700 (2008)
18. S.-L. Jang, C.-C. Shih, C.-C. Liu, M.-H. Juang, A0.18 mCMOS quadrature VCO using the quadruple push-push technique. IEEE Microw. Wirel. Compon. Lett. **20**(6), 343–345 (2010)
19. S. Zafar, M. Awan, T.Z.A. Zulkifli, 5-GHz low-phase noise quadrature VCO in 0.13 um RF CMOS process technology, in *Procrrdings of IEEE 10th Annual Wireless and Microwave Technology Conference*, 2009, pp. 1–4
20. T.F. Chan, H.C. Luong, A 0.8 V CMOS quadrature LC VCO using capacitive coupling, in *Proceedings of IEEE Asian Solid-State Circuits Conference*, 2007, pp. 436–439

Chapter 10
Cyber Situational Awareness for CPS, 5G and IoT

Elizabeth Chang, Florian Gottwalt and Yu Zhang

Abstract The 2020 wireless strategy is centred on creative 5G and IoT with the front runners from US, EU, China, Japan and Korea. There are trillions of dollars invested towards 2020. The wireless future is designated to be mobile. While the confidence of the future mobile technology will help innovate our government, workforce, industry and social media, the key threat to future mobile wireless development is the growing concern of its security and privacy through anomaly detection. This keynote presents key features of 2020 wireless strategy and new approaches on cyber situation awareness—the enablement to achieve 2020 wireless leadership.

Keywords 2020 wireless strategy · 5G · IoT · Cyber situation awareness · Natural law · Anomaly detection ontology

1 Introduction

In recent years, leadership in future mobile and frontier wireless started in 2014 with White House $450 million grants to upgrade wireless industry (Degrasse 2014), continuing into 2015 with billions of funds invested in 5G by Europe (Fioretti and Abboud 2015). This further extended with trillion dollar efforts to step

---

Invited Plenary for ICNETS2 2017.

E. Chang (✉) · F. Gottwalt · Y. Zhang
University of New South Wales and Australian Defence Force Academy,
Canberra, Australia
e-mail: e.chang@unsw.edu.au

F. Gottwalt
e-mail: f.gottwalt@adfa.edu.au

Y. Zhang
e-mail: y.zhang@adfa.edu.au

S.R.S. Prabaharan et al. (eds.), *Frontiers in Electronic Technologies*,
Lecture Notes in Electrical Engineering 433, DOI 10.1007/978-981-10-4235-5_10

up the pace of 5G development in China, Japan and Korea (Huawei 2015). We can already see wide-spread acknowledgement from the consumers, governments, industries and social media.

The key threats to future wireless networks, namely 5G and IoT, adoption is the concerns of their security and privacy. Traditional cyber security methods need to continue their efforts to address the debilitating inability to keep pace with unknown attacks, which leads to high concerns for future wireless technology use beyond social and entertainment.

2 2020 Wireless Strategy: 5G and IoT

Future wireless strategy is set to be the next generation of mobile telecommunication development. While work has been done in Cyber-Physical Systems, Web-of-Things and Clouds, the frontier will be 5G and IoT. The 2020 wireless strategy will include a collection of the current wireless communication and pre-existing wireless technologies covering cellular, Wi-Fi, RFID, Bluetooth and others to allow higher coverage and availability, and higher network density in terms of cells and devices. The key differentiator will be greater connectivity as an enabler for Machine-to-Machine (M2M) services and ultimate Internet of Things (IoT).

2.1 From 1G, 2G, 3G, 4G to 5G

The quantum leap of each generation of mobile technology is the demand to improve connectivity, mobile experience, smart-phones and remote control of objects from anywhere in the world. Table 1 demonstrates the evolution from 1G to 5G.

5G targets massive connection, which integrates all modern communication concepts and technologies including Milliwave, Internet of Things, Wireless Sensor Network (WSN) and M2M (GSMA 2014).

The requirement for 5G to fulfill IoT is highly rigorous. There is a great deal of importance for 5G to enable a massive number of devices to be connected simultaneously to the network, and support all-time connected cloud services, as well as more machine type devices for IoT (DOCOMO 2014) with very high throughput services. As a fundamental part of the 5G era, the IoT will lead to more connected devices and also new use cases for small cell deployments (5G-PPP 2014). It will also help revolutionise the way we communicate by supporting immersive applications that demand exceptionally high-speed wire-less connections (NOKIA 2015). 5G mobile communication will bring a full scale IoT to reality (Samsung 2015).

The advent of the 5G concept and its connectivity technology, which is expected to be ubiquitous, reliable, scalable and cost-efficient, is considered as a potential key driver for the yet to emerge global IoT (Palattella et al. 2016).

Table 1 Evolution from 1G, 2G, 3G, 4G to 5G

Generation	Primary services	Key differentiator	Challenges
1G	Analogue phone calls	Mobility	Poor spectral efficiency, major security issues
2G	Digital phone call and messaging	Secure, mass adoption	Limited data rates—difficult to support demand for internet/e-mail
3G	Data	Better internet experience	Real performance failed to match hype, failure of WAP for internet access
3.5G	Broadband data	Broadband internet, applications	Tied to legacy, mobile specific architecture and protocols
4G	All-IP services (including voice, messaging)	Faster broadband internet, lower latency	Quality of Service drop and delay
5G	Control smart devices anywhere anytime	Massive Connection, full scale IoT, all-time cloud services	Anomaly Detection

In addition, 5G features include increased data rate, reduced end-to-end latency, improved coverage (compared to 3G and 4G), and potentially have the ability to support even the most demanding IoT applications in terms of communication requirements. As 5G technology research and development continues, the 5G mobile broadband requirements and standards are growing to maturity. Its building block, the IoT architecture, is moving forward to become reality.

2.2 5G Features—The Wireless Frontier

5G is far beyond 4G and the IoT. Its new hardware and software enablers will allow 5G to be significantly different from any types of current wireless communications, including frequency band, data rate, network topology, architecture, enablers, routing type and data transmission—this is shown in Table 2.

The key significance of 5G lies on its network architecture and data transmission.

In 5G network, the architecture will be a Software-Defined Network Functionality Virtualisation (NFV) enabled network architecture supported by the emerging enablers, i.e. SDN and NFV. The Software-Defined Networking (SDN) paradigm, initially designed for wired networks, has recently gained a lot of interest in the wireless environment, (Granelli, Gebremariam et al. 2015) and been discussed widely in the IoT applications (Valdivieso Caraguay, Benito Peral et al. 2014). Today, SDN techniques have been seen as promising enablers for 5G carrier networks (Hakiri and Berthou 2015) to address the potential relevance between frequency spectrum and network information (Cho, Lai et al. 2014). On the other

Table 2 5G features and contrast to other existing wireless networks

	Hardware						Software						Limitation
	Network type	Technology	Reference standard	Frequency bands	Data rate	Max distance	Network topology	Architecture	Network enabler	Data routing		Transmission	
IoT	Wired network	Ethernet, XDIS	IEEE 802.3 u/z	Up to 2.2 MHz	Various, Up to 100 fibcs	Up to 50–70 km, 5.4–1.3 km	Centralised	Client-Server Architecture					Shared medium, Physical connection
	Wireless network (except cellular network)	wta RFID, Ziflbee, Bluetooth, Satellite	IEEE 802.11 a/b/g/n, IEEE 802.16 a/d/e/m, IEEE 802.15	2.4 GHz, 5 GHz, 2–66 GHz, 900–1800 MHz, 1900–2100 MHz, 800–2600 MHz, C band, Ku band, Ka band	Various, up to 600 Mbps	Up to 100 m, up to 50–80 km, up to 100 km	Partially distributed	Service-Oriented Architecture (SOA)	Cloud computing, D2D communication	Single-channel	LEACH, HEED, AODV, VCCSR, BGR	Limited cross-network transmission	Sensitive to radio interference, Low practical data rate, High installation and operational costs, Limited wireless spectrum, Low traffic capacity, Limited node computation
3G	Cellular network	HSPA+, CDMA	GSM, 3GPP	1.8–2.5 GHz	Up to 30 Mbps	Regional (country wide)	Distributed	UMTS network architecture	User Equipment (UE), Radio Access Network (RAN)	Base station radio broadcast		Broadcast	Limited data rate, Tied to legacy
4G	Cellular network	LTE, WiMAX	3GPP, IEEE	2–8 GHz	Up to 1 GbRS	Regional (country wide)	Distributed	Flat LTE network architecture (All-IP architecture)	All-IP structure, enhanced nodes feNosMD	Base station radio broadcast, WiMAX		Broadcast	QeS Drop and Delay
5G	5G network (network integration)	5G	N/A	2.3 GHz, 2.6 GHz, 5.25 GHz, 26.4 GHz, 58.68 GHz, Up to 70 GHz	Up to 100 fibcs	Global	Fully distributed	Software-defined NFV enabled network architecture	SDN, RAT, NFV, Cloud service, millimetre wave technology	Multi-channel		Multi-cast, multi-radio, multi-hop, multi-RAT, multi-channel	Anomaly Detection

hand, Network Function Virtualisation (NFV) is a complementary technology of SDN, destined to impact future 5G networks. NFV aims to virtualise a set of network functions, by deploying them into software packages, which can be assembled and chained to create the same services provided by legacy networks. By moving network functions from dedicated hardware into general purpose computing/storage platforms (e.g. servers), NFV technologies will help manage heterogeneous devices. Moreover, by implementing the network functions in software packages that can be deployed in virtualised infrastructure, NFV offers scalability and large flexibility in operating and managing mobile devices.

In regards to data transmission, cellular networks deploy communication depending on Base Station (BS) and Relay Station (RS) broadcasting radio signals to the mobile subscribers: even WiMAX radio transmission is based on this method. In the IoT network environment, the type of data transmission is various, among which the most famous and promising one is multi-hop transmission. It allows nodes in a distributed network to communicate with each other and to cooperatively pass their data through the network to the destination without relying on a pre-existing infrastructure, such as routers in wired networks or access points in managed (infrastructure) wireless networks. Instead, each node participates in routing by forwarding data for other nodes, so the determination of which nodes forward data is made dynamically on the basis of network connectivity.

Data transmission in 5G networks is more complicated than that in current wireless networks, as heterogeneous devices are to be involved in 5G which and exchange data are constraint to certain radio frequency. In addition, smart devices will be equipped with hardware and software module supporting multi-Radio Access Technology (multi-RAT) and in 5G environment, various range of radio frequency access must be switched on the smart devices based on changing radio environment or different use demand. Unlike most current single-channel data routing method in wireless network, there will be massive data routing based on multi-channel using multiple radios to reduce interference during the communication among nodes.

3 Cyber Situation Awareness

Cyber Situation Awareness focuses on anomaly detection and incident response, and is a multifaceted field of study. From a technical point of view, situation awareness is about compiling, processing and fusing different data sources, with the core technology being data fusion. From a cognitive viewpoint, situation awareness is the human capability to perceive and comprehend data as well as draw conclusions from it [1].

The research in this area supports the future of 5G and IoT in the following ways:

- Advise cyber security through big data fusion
- Assist cyber instruction detection approaches
- Discover context to activities for decision making
- Alert importance and impact of event
- Draw conclusions about anomaly behaviour
- Predict and forecast future steps of adversary
- Reduce the cognitive load for human analyst
- Enhance visualisation for human perception

4 Key Challenges

While it would be desirable to have a Cyber Situation Awareness system that supports human analysts and anomaly detection, there is no ready used solution in the market and requires more research to be done for the realisation of the future mobile environment. While here are plenty higher level reasoning frameworks, key technical issues to address security and privacy are yet solved and no proven technology is available.

For example, one technical issue to be solved is how to enhance the effectiveness of anomaly detection sensors. Current anomaly detection algorithms proposed in the literature have rarely found their way into real-world applications as they generally only perform well in laboratory environments and not in complex, dynamic real networks. Commonly this leads to a huge amount of false alarms resulting in an overwhelming cognitive load for security analysts. Furthermore, the developed algorithms lack in providing reasoning information, meaning their output is trivial.

Another challenge and way to improve anomaly detection methods is to integrate more contextual information. Researchers commonly treated all input attributes of an anomaly detection method in the same way without considering contextual meanings. Only reasonable work has incorporated combined contextual categories in the detection algorithm and to the best of our knowledge, none has considered all contextual dimensions to model the environment.

While it is often mentioned in the literature that the fusion and integration of more, heterogeneous data sources could reduce the amount of false alarms and enhance inferring capabilities, only reasonable amount of work has evaluated applied data fusion on real-world scenarios. Furthermore, the approaches proposed lack in comparing the cost/benefit efficiency of adding more data. This is particularly important considering the combination of various data sources can lead to huge computational challenges. Nevertheless, only little work has taken these challenges into account and more scalable solutions are required.

While it is the vision to have a cyber situation awareness system which covers all these aspects, in a self-aware, self-learning and self-protecting manner without the required interaction of a human, the reality is far from it [2]. One major reason is that traditional approaches are only partly applicable to the cyber domain due to significant differences in the physical and cyber world [3]. In comparison to traditional situation assessments in the physical environment, e.g. piloting an airplane, the cyberspace is massive in size and with that much harder to comprehend [3]. Furthermore, to be able to understand it, knowledge from multiple disciplines is required.

Another challenge to obtain situation awareness in the cyber domain is uncertainty. The correctness of sensor data in the cyber environment such as produced by an IDS is not reliable and in addition, contextual cues are incomplete and uncertain, making it hard to find the motivation and capabilities of adversaries [4].

To address these issues, several frameworks [3, 5, 6] have been proposed, yet core technical challenges need to be solved before higher levels of SA such as reasoning of events or prediction of future situations can be achieved [2]. Two key technical issues for that are to increase the effectiveness of cyber sensors and to develop intelligent solutions which correlate and fuse several data sources into one unified logical representation. To address these issues, the next section discusses challenges intrusion detection sensors are facing and how they can be enhanced.

5 Anomaly Detection for Future Wireless Communication

A lot of research has been done on anomaly detection for intrusion detection, and the current state of the art approaches are utilising statistical methods, machine learning methods and data mining methods [7]. Anomaly detection techniques can be divided into point anomalies, contextual anomalies (Activity, time, location, relations, etc.) and collective anomalies [8]. **Point anomalies** are anomalies where individual records are structurally different from the rest of the dataset. **Contextual anomalies** are records which are anomalous in a specific context but not without this context. An example for that is a credit card transaction of $10 from an institution in a foreign country. Assuming a transaction of $10 is nothing uncommon, the context (location) of it makes it anomalous. **Collective anomalies** describe a collection of instances which are anomalous joint together but not individually [8]. The majority of recent studies are focusing on the increase of accuracy of point anomaly detection techniques, and although researchers have obtained excellent detection accuracies on IDS datasets, these techniques rarely have found their way into actual deployments [9, 10].

One main reason for this is it is extremely hard to differentiate between normal and anomalous behaviour in the cyber world. To define what is expected to be normal, anomaly detection techniques are usually analysing traffic over a certain time frame (training phase) and generate a baseline out of it. This is, however, not very efficient in an environment where the nature of normal and anomalous behaviour is changing over time in a quick and unpredictable fashion [8]. As a consequence, generally large amounts of false alarms are generated.

Besides the high false-alarm rate issue, the reasons for that can also be found in the nature of data mining and machine learning techniques. Machine learning and data mining algorithms were originally developed with the intention to detect similar behaviour, rather than anomalies and in addition learning techniques are only restricted useable in a fast changing environment like the cyberspace [9, 11].

A high volume of anomaly detection research has neglected the computational complexity of data mining and machine learning techniques. As described in [10], the majority of approaches are not taking the computational effort into account and considering the immensely growing cyberspace this is a huge challenge which needs to be addressed. Zuech et al. have provided a good survey on IDS in times of Big Data and how the fusion of big heterogeneous data sources could enhance the detection performance. Nevertheless, they have also concluded that it is essential to assess whether the cost to fuse several sources is worth the gain, as even a single data source can face big data challenges [12].

After an event has been detected the next step for the human analyst is to interpret it on validity and criticality. The output of anomaly detection techniques, however, makes it hard to be interpreted as very often only rudimentary results are returned. Commonly the suspicious values are displayed without any reasoning which leads to a semantic gap between the results and their interpretation [9].

Even if all these issues would be solved in an academic context, it would still not be guaranteed that these approaches would be applicable in a real deployment. Another huge challenge is that most of the results achieved by researchers would not be representative in an actual deployment, simply due to the complexity of a real network and the lack of availability of realistic intrusion detection datasets for researchers [11, 12].

In fact, it has been observed that the complex problem of intrusion detection has been treated too general by many researchers [9]. Rather than focusing on one specific problem (attack type), very often they have evaluated one technique for the detection of any kind of attack. Looking at the characteristics of the cyberspace and attacks, this is, however, the wrong approach as there cannot be one "ultimate" individual solution to cover the whole spectrum and variety of attacks. Rather than searching for that "ultimate" solution, it is required that several detection strategies are combined to address the complexity of the problem.

A technique as proposed in our most recent publication [13] could be utilised in this combined fashion or other approaches such as the one described by Veeramachaneni et al. should be the orientation, where different statistical, data mining and machine learning techniques are combined in a solution for a scalable real-world application rather than an optimal laboratory solution [14].

Further potential to enhance the effectiveness of anomaly detection algorithms lies in considering more contextual information and combinations of other heterogeneous data sources for which previous work is outlined in the next section and Sect. 2.4 [2, 15].

Multi-sensor data fusion is "*a process dealing with the association, correlation and combination of data and information from single and multiple sources to achieve refined position and identity estimates, and complete and timely assessments of situations and threats as well as their significance*" [16]. We adopt this definition and consider data fusion as a collection of technical means to combine various data sources in the SA process with the ultimate goal to gain SA.

A lot of these approaches have delivered promising results and in fact, combining multiple data sources has been suggested as being key to detecting hard detectable attacks such as Advanced Persistent Threats (APT) attacks [17, 18]. Nevertheless, the results have to be treated with caution. Most of the approaches have been evaluated in laboratory environments using datasets which are not reflecting the real world such as the KDD99 dataset [9]. Furthermore, it is surprising that the results were never compared from a cost/improvement perspective. Adding additional data sources to obtain better results can be very costly and it has to be evaluated whether the computational effort is worth the enhancement.

6 A Cyber Situation Awareness Framework for Anomaly Detection

Work led by author Gottwalt of UNSW [19, 20] using Natural laws for anomaly detection have shown the potential to detect anomalies in other areas and have the advantage to neither require a training phase nor large computing power. With the assumption that normal traffic obeys a natural law and anomalous behaviour deviates from it, several TCP flow characteristics have been evaluated against their compliance with four natural laws for normal traffic firstly. The results have shown that almost all characteristics do have a connection to the natural laws with some following them very accurately.

Secondly, the same tests were performed on several datasets containing normal and anomalous traffic to see whether the laws are still present or not and if not, if this can be used to detect anomalies. The results have demonstrated that the characteristics deviate significantly from the laws when anomalies are present. Furthermore, outstanding detection results have been obtained on several intrusion detection datasets, which highlights the enormous potential of the approach.

One of the key issue addressed in this techniques is, after being able to process this high volume of data, how to get insights out of it. Conventional methods, which are very often based on deterministic rules, start to struggle when it comes to a certain point of complexity and volume. The approach is based on the natural law to identify similarities and detect abnormalities in high-volume data sets. While much of research done in applying data mining techniques to detect intrusions, few focus on data summarisation and visualisation of events. Our Framework to address this is shown in Fig. 1.

To evaluate the applicability of the concept, the outcome of the proposed algorithm is based on Hoplaros et al., which are conciseness, information loss, intelligibility and interestingness. The data set used is composed of firewall logs, which contain information structured in key-value pairs. The logs are comprised of

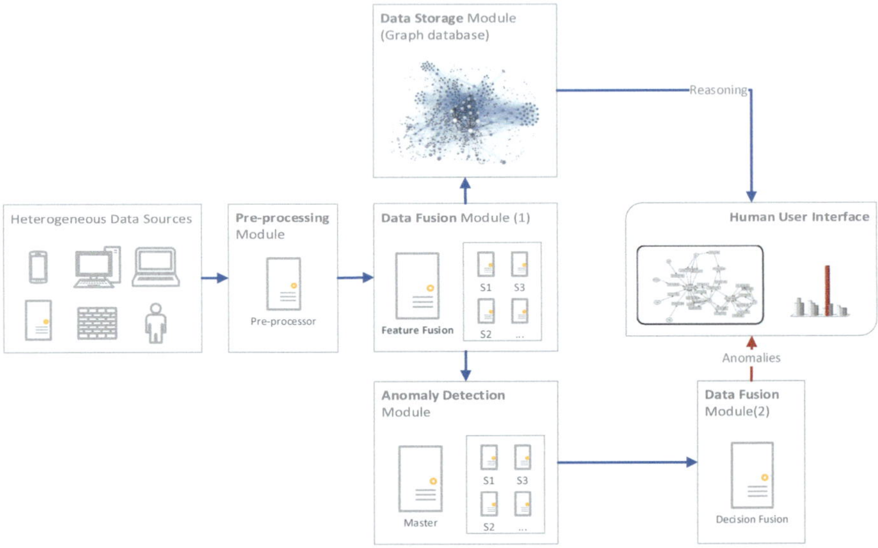

Fig. 1 The anomaly detection for cyber situation awareness for future wireless networks [20]

Table 3 Anomaly detection test run [19]

Threshold	0.01%	0.10%	1%	10%
Amount of clusters	1209	216	1	0
Amount of cluster members	6917	7219	2317	0
Amount of outliers	9237	8935	13837	16154
Amount to analyse	10446	9151	13844	16154
Conciseness	15.46	17.65	11.67	10
Information loss	41	43	24	0
Intelligibility	11	16	27.7	0

general information such as time, device name and type of event and other relevant attributes like, e.g. source IP address, source IP port, destination IP address, destination Port, Protocol, Number of packets and bytes as well as the TCP flags (Table 3).

The approach is light weight, low computational cost and suitable for future generation of wireless computation. The obtained results indicate that the proposed framework is generally applicable with the achievement of generating a good overview about all occurring events in a defined time frame. It includes a compression techniques the input data to a tenth of the size with a reasonably low information loss, indicates it could be adapted to large complete network and with adjusted visualisation techniques with an easy interpretable result for humans.

7 Anomaly Detection Ontology for Cyber Situation Awareness

Building ontology for wireless network anomaly detection is an emerging method to achieve high accuracy, comprehensive coverage, self-organisation and flexibility for future wireless network including 5G and IoT. We leverage the power of Natural Language Processing (NLP) and Crowdsourcing for this purpose by constructing an automatic framework which aims at developing ontology of anomaly detection with case study on Scopus documents. Its aims to build an automatic method to ensure that reliable knowledge in the form of ontology are extracted and connected about anomaly detection. The heuristic method has been proposed which reduce the search space substantially which makes the proposed framework agile, low cost of computation and efficient.

The anomaly detection ontology consists of three integrated modules including *Information Exploration Module*, *Relationship Construction Module* and *Verification Module*. Figure 2 presents the workflow of these modules.

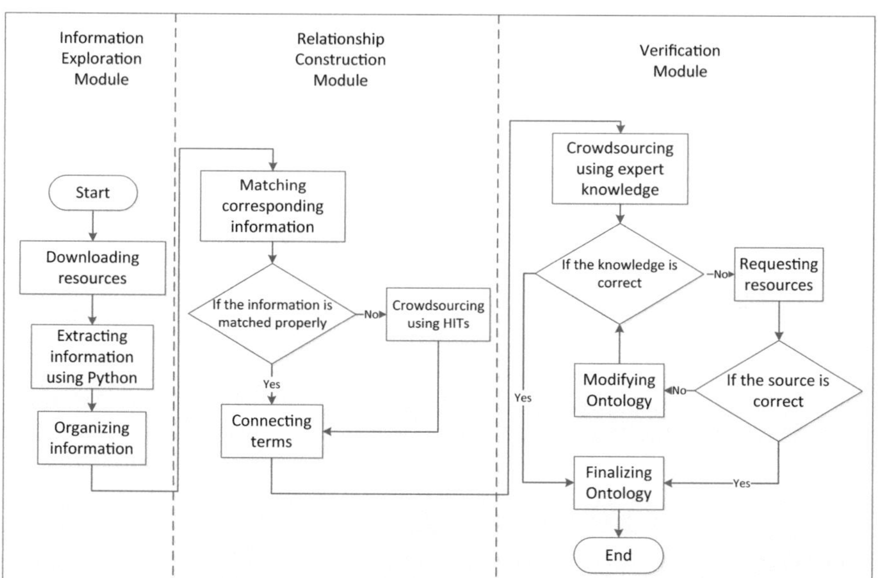

Fig. 2 The automated intrusion detection ontology generation

7.1 *Information Exploration Module*

In the information exploration module, intrusion detection related information is extracted from Scopus which covers prestigious journals and conferences in various fields specially engineering. We firstly screen academic papers with high quality using the functions provided by Scopus so that the papers with more citations, from better journal or conference sources, or written by authors with higher reputation will be filtered out. This step is important as it can help avoid poorly constructed papers which reduces the uncertainty during the developing process and increases the accuracy of the formed ontology. Secondly we extract the information including intrusion types and counterattack techniques due to importance of thoroughly digging out all the basic knowledge regarding to intrusions and detecting methods for intrusion detection ontology development. In order to approach this, NLP is proposed to explore the useful information from the downloaded articles. Python and Natural Language Toolkit programming language are used in order to extract all the nouns in front of the term "attack" and all the sentences with the term "propose" or "present" inside. The extracted nouns represent different types of attacks while the sentences contain the counterattack techniques used in different resources. Then the extracted information will be organised and saved in database.

7.2 *Relationship Construction Module*

After acquiring various *types of intrusions* and *counterattack techniques*, the relationship construction module aims to connect them individually to build up the relationship map for intrusion detection knowledge. A "Step-by-Step" model is proposed to match the corresponding information extracted from the resources, which means this model will connect the intrusions and techniques that were proposed to address the corresponding intrusions.

In the "Step-by-Step" model, the titles of all the academic articles are firstly given as input, so that the articles with *intrusion type* and *technique* which is proposed by the author to address the intrusion type mentioned together in titles will be flagged. In this case, intelligently the search space of framework is reduced substantially. If title itself is not representative of the context intention regarding relationship between *types of intrusions* and *counterattack techniques*, paper's abstract is analysed. If the abstract assists us in building a relationship between *types of intrusions* and *counterattack techniques*, we stop the searching process and insert the new relation in the ontology model, otherwise introduction and conclusion parts of the paper are analysed. Generally, there are three possibilities in this space when we are focusing on papers' conclusion: (1) show nothing, (2) show only a *proposed technique* but no *intrusion type*, (3) show several *intrusion types* and a *proposed technique*. Papers with this feature still show no information are removed from search space when did not help us in the ontology development.

The second type is marked as general techniques and the third one is filtered and needs further process. In the third type, there are two general possible cases: (1) one of the several extracted *intrusion types* appears more frequently than others, (2) these several *intrusion types* show similar occurrence frequencies. The first possible result indicates that the *proposed technique* is related to that *intrusion type* appearing more times than others, because this *intrusion type* is mentioned more frequently in the introduction and conclusion parts of the paper. When it comes to the second possible result, similar occurrence frequencies of the extracted *intrusion types* make it difficult for computers to decide which one(s) is supposed to be filtered, so this type of result will be sent to the crowd workers to determine appropriate relationship in this situation using predesigned HITs.

This "Step-by-Step" model was designed based on real experiment. Figure 3 presents an example in which 46 articles were randomly chosen as database. Out of these 46 inputs, 5 outputs showed intrusions and corresponding techniques extracted from titles, 8 outputs from abstracts. Then out of the rest 33 inputs, 3

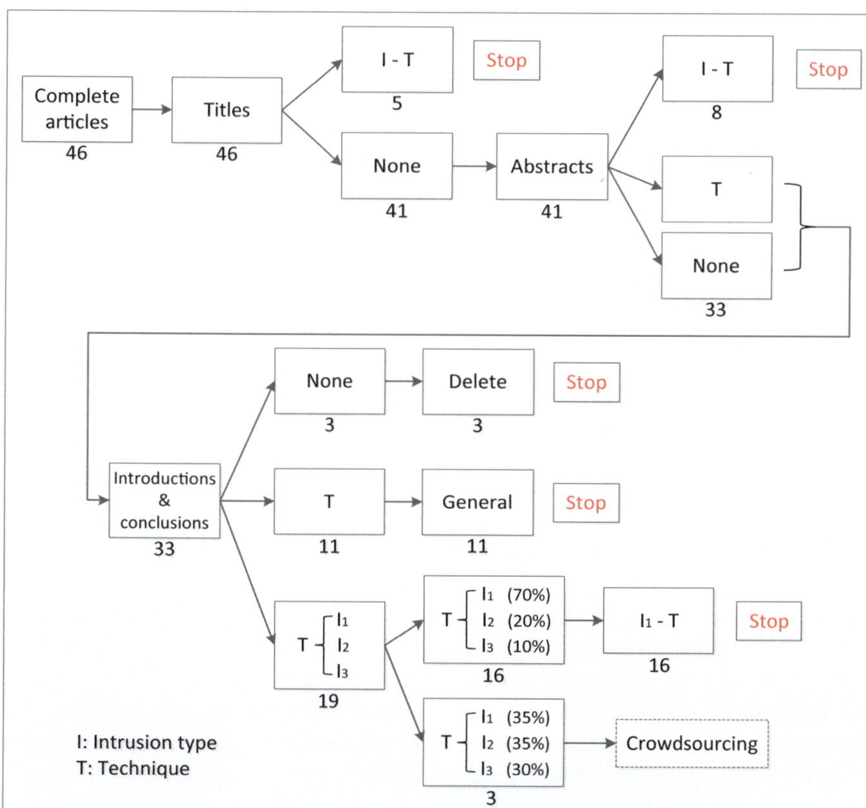

Fig. 3 The step-by-step abstraction of all the concepts and techniques for anomaly detection

outputs were deleted as they still did not show any information, 11 outputs were marked as general techniques which were proposed to address many types of intrusions, 19 outputs showed one technique and several intrusion types. In the 19 outputs, 16 of them showed that one of the intrusion types appears much more frequently than other types, only 3 out of 19 outputs indicated types of intrusions with similar occurrence frequencies, and needed further crowdsourcing process.

7.3 Verification Module

In this module, the intrusion detection ontology is evaluated using the crowdsourcing technique. Specifically, the developed ontology and experiment results can be accessed to domain experts or the professionals in intrusion detection field so that they are capable of verifying whether the terms and connections in the ontology are correct. Firstly, the ontology will be completely accessed to the crowdsourcing experts. If the experts have doubt about any terms or relationships in the ontology, the processing details in the "Step-by-Step" model for each article as well as the source articles will be provided for them to refer to and determine the relationships, otherwise the ontology will be finalised. If any of the relationships in the ontology is found as incorrect or the experts are confident to indicate the errors in the ontology, they are allowed to leave feedbacks such as comments, suggestions and references. After collecting and analysing feedbacks, the ontology would be modified periodically based on the source document and the provided feedback, and finalised afterwards. In this way, the mistakes in both the ontology and the program can be located and revised using the intelligence of the crowdsourcing method.

The approach incorporates a crowdsourcing and NLP based auto intrusion detection ontology generation. The ontological concepts are extracted through source articles using NLP program and the ontology is verified by crowdsourcing methods. This leads to the ontology generation and verification. A preliminary prototype tool has been produced.

8 Conclusion

The 2020 wireless strategy are centred on 5G and IoT, with the key threat of security and privacy. We presented special features of 2020 5G and IoT, the key challenge being anomaly detection for security and privacy. We have also presented new approaches of Cyber Situation Awareness through anomaly detection and anomaly ontology development address the weakness of existing cyber security research and that could be used for future wireless technology development.

References

1. U. Franke, J. Brynielsson, Cyber situational awareness—a systematic review of the literature. Comput. Secur. **46**, 18–31 (2014)
2. P. Barford et al., Cyber SA: situational awareness for cyber defense, in *Cyber Situational Awareness* (Springer, 2010), pp. 3–13
3. M. Eric, T. Michael, An alternative framework for research on situational awareness in computer network defense, in *Situational Awareness in Computer Network Defense: Principles, Methods and Applications*, ed. by O. Cyril and O. Thomas (IGI Global, Hershey, PA, USA, 2012), pp. 71–85
4. J. Yen et al., RPD-based hypothesis reasoning for cyber situation awareness, in *Cyber Situational Awareness* (Springer, 2010), pp. 39–49
5. M. Sudit, A. Stotz, M. Holender, Situational awareness of a coordinated cyber attack, in *Defense and Security* (International Society for Optics and Photonics, 2005)
6. A.C. Squicciarini et al., Situational awareness through reasoning on network incidents, in *Proceedings of the 4th ACM Conference on Data and Application Security and Privacy* (ACM, San Antonio, Texas, USA, 2014), pp. 111–122
7. C.A. Catania, C.G. Garino, Automatic network intrusion detection: current techniques and open issues. Comput. Electr. Eng. **38**(5), 1062–1072 (2012)
8. V. Chandola, A. Banerjee, V. Kumar, Anomaly detection: a survey. ACM Comput. Surv. (CSUR) **41**(3), 15 (2009)
9. R. Sommer, V. Paxson, Outside the closed world: on using machine learning for network intrusion detection, *in 2010 IEEE Symposium on Security and Privacy* (2010), pp. 305–316
10. A. Milenkoski et al., Evaluating computer intrusion detection systems: a survey of common practices. ACM Comput. Surv. **48**(1), 1–41 (2015)
11. F. Gottwalt, A.P. Karduck, SIM in light of big data, in *2015 11th International Conference on Innovations in Information Technology (IIT)* (IEEE, 2015)
12. R. Zuech, T.M. Khoshgoftaar, R. Wald, Intrusion detection and big heterogeneous data: a survey. J. Big Data **2**(1), 1–41 (2015)
13. F. Gottwalt, A. Waller, W. Liu, Natural laws as a baseline for network anomaly detection, in *15th IEEE International Conference on Trust, Security and Privacy in Computing and Communications (IEEE TrustCom-16)* (IEEE CPS, Tianjin, 2016)
14. K. Veeramachaneni, I. Arnaldo, A. Cuesta-Infante, V. Korrapati, C. Bassias, K. Li, AI2: training a big data machine to defend, in *The 2nd IEEE International Conference on Big Data Security* (2016)
15. S. Avidan, Spatialboost: adding spatial reasoning to adaboost, in *European Conference on Computer Vision* (Springer, 2006)
16. F.E. White, Data fusion Lexicon, DTIC Document, 1991
17. P. Giura, W. Wang. A context-based detection framework for advanced persistent threats, in *2012 International Conference on Cyber Security (CyberSecurity)* (2012)
18. J.D. Vries et al., Systems for detecting advanced persistent threats: a development roadmap using intelligent data analysis, in *2012 International Conference on Cyber Security (CyberSecurity)* (2012)
19. F. Gottwalt, A. Karduck, SIM in light of big data, in *IEEE TrustCon 2015*, China
20. F. Gottwalt, *A Scalable Context-Awareness Anomaly Ressoning—A SCALAR Approach* (UNSW, 2016)

MIX
Papier aus verantwortungsvollen Quellen
Paper from responsible sources
FSC® C105338

If you have any concerns about our products,
you can contact us on
ProductSafety@springernature.com

In case Publisher is established outside the EU,
the EU authorized representative is:
Springer Nature Customer Service Center GmbH
Europaplatz 3, 69115 Heidelberg, Germany

Printed by Libri Plureos GmbH
in Hamburg, Germany